FLIGHT PHYSICS - MODELS, TECHNIQUES AND TECHNOLOGIES

Edited by **Konstantin Volkov**

Flight Physics - Models, Techniques and Technologies
http://dx.doi.org/10.5772/intechopen.68297
Edited by Konstantin Volkov

Contributors

David Johnson, Mudassir Lone, Vilius Portapas, Gaetan Dussart, Alessandro Pontillo, Constantin Rotaru, Michael Todorov, Illia Stanislavovych Kryvokhatko, Oleksandr Masko, Joel Guerrero, Jiri Cecrdle, Mohamad Taha, Jozsef Rohacs, Youjin Kim, Ali Al-Abadi, Antonio Delgado, Haibo Dong, Ayodeji Bode-Oke, Chengyu Li, Wei Ding

Published by InTech

Janeza Trdine 9, 51000 Rijeka, Croatia

Notice

Statements and opinions expressed in the chapters are these of the individual contributors and not necessarily those of the editors or publisher. No responsibility is accepted for the accuracy of information contained in the published chapters. The publisher assumes no responsibility for any damage or injury to persons or property arising out of the use of any materials, instructions, methods or ideas contained in the book.

Publishing Process Manager Kristina Kardum
Technical Editor SPi Global
Cover InTech Design team

First published February, 2018
Printed in Croatia
Legal deposit, Croatia: National and University Library in Zagreb

Additional hard copies can be obtained from orders@intechopen.com

Flight Physics - Models, Techniques and Technologies, Edited by Konstantin Volkov
p. cm.
Print ISBN 978-953-51-3807-5
Online ISBN 978-953-51-3808-2

For EU product safety concerns:
IN TECH d.o.o., Prolaz Marije Krucifikse Kozulić 3, 51000 Rijeka, Croatia,
info@intechopen.com or visit our website at intechopen.com.

Meet the editor

Dr. Volkov is a senior lecturer in thermofluids at the Kingston University (London, UK). He holds a PhD degree in fluid mechanics. After completion of his PhD, Dr. Volkov worked at the Baltic State Technical University (Russia), University of Central Lancashire (UK), and University of Surrey (UK). His areas of expertise cover multidisciplinary areas from design and optimization of energy systems to fundamental problems focused on modeling and simulation of turbulent multiphase flows. He is the chartered engineer and a member of Institute of Physics, Institution of Mechanical Engineers and Combustion Institute. He is the author of more than 120 scientific papers and a member of editorial board and scientific committee of a number of journals and conferences.

Contents

Preface

Flight physics is a complex and interdisciplinary discipline, combining aerodynamics, fluid and solid mechanics, flight control and mechanics, and other scientific disciplines. In terms of aerodynamics, the study of flight combines wind tunnel testing, similarity scaling principles, analytical modeling, computational fluid dynamics (CFD) and empirical test data.

Aircraft development is triggered by the success of past years' research and development activities resulting in increased lightweight performance of the aircraft, by using new materials and advanced manufacturing and design methods. Practical demands led to more complex aircraft architectures and consequently a need for higher efficiency and hence less conservative designs.

The dynamic behavior of aircraft is becoming more and more a design-driving topic. Considerable changes of aircraft design processes lead to significant reduction of development times, while including a number of disciplines in the early phases of design activities in order to find an optimum design. One of the main objectives in the aerospace industry is the reduction of aircraft development time and provision of robust solutions with improved quality. Simulation of the dynamic behavior is an important part of the design process and has a strong impact on safety and comfort. However, description and simulation of the dynamic behavior are also a challenging topic, since dynamic simulation requires detailed and advanced mathematical models. Improved simulation capabilities have a rather large impact on improving cost efficiency both with respect to aircraft development cost and aircraft operational cost.

The book offers a general overview of the concepts, theories and models underlying flight physics. The book focuses on the synthesis of the fundamental disciplines and practical applications involved in the investigation, description, and analysis of aircraft flight including applied aerodynamics, aircraft propulsion, flight performance, stability, and control. The book covers the aerodynamic models that describe the forces and moments on maneuvering aircraft, and provides an overview of the concepts and methods used in flight dynamics. A particular focus of the book is development of theoretical, computational and experimental methods in aerodynamics. The book covers some topics in depth while offering introductions and summaries of others. There are 11 chapters in the book covering different aspects related to flight physics.

Learning from Nature: Unsteady Flow Physics in Bioinspired Flapping Flight

The aerodynamics of free-flying insects is studied based on the approach combining high-speed photogrammetry, three-dimensional surface reconstruction, and computational fluid dynamics. The proposed approach is capable of measuring the wing kinematics, and surface deformation and simulating its aerodynamic performance. Four different species of insects (dragonfly, damselfly, butterfly, and cicada) are investigated to demonstrate the possibilities of the approach. A deep understanding of flapping wing mechanism, revealing the underlying flow physics for future designs and applications, is achieved.

Helicopter Flight Physics

The principles that constitute the fundamentals of helicopter flight physics, starting from the basics of the main rotor aerodynamics and of the component parts related to flight control, are considered. A short history of helicopter development is also presented. The main constructive solutions for helicopters and the basic equations of fluid mechanics are applied on a helicopter model with one main rotor and tail rotor. The ground effect, autorotation, stability and helicopter control are discussed, and the main factors that determine the helicopter performances are mentioned.

Flight Dynamic Modeling and Simulation of Large Flexible Aircraft

The approach to modeling and simulation of large flexible aircrafts is presented. The approach requires a careful integration of aerodynamic models with models for structural dynamics. The coupled aero-structural model then provides forces and moments to the equations of motion. Two cases are used to demonstrate possibilities of the proposed methodology. The first case focuses on aircraft response to a gust that has a spanwise varying profile, and the second case investigates aircraft dynamics during control surface failure scenarios.

Aerodynamic Characteristics and Longitudinal Stability of Tube-Launched Tandem-Scheme UAV

Tube-launched unmanned aerial vehicles (UAVs) are often implemented with aerodynamic scheme with forward and rear wings (tandem scheme). UAV aerodynamic characteristics for different wing rotation angles are discussed. The results show that UAV is unstable with wings' rotation angles up to 60 degrees because rear wings produce lift ahead of center of gravity. For large angles of wing rotation (low sweep angles), UAV model is stable in wide range of angles of attack.

Airfoil Boundary Layer Optimization toward Aerodynamic Efficiency of Wind Turbines

An airfoil optimization method for higher aerodynamic efficiency and laminar boundary layer enlargement is discussed, and advantages of laminar boundary layer expansion in airfoil of horizontal axis wind turbine blades are presented. The optimized airfoil shows enlarged laminar boundary layer region in all flow regime with a higher aerodynamic efficiency and the increased gliding ratio. The smaller laminar separation region and reduced stall regime of CFD simulations illustrate the desirable aerodynamics of the resulted airfoil.

LES of Unsteady Aerodynamic Forces on a Long-Span Curved Roof

Unsteady aerodynamic forceson long-span curved roofs with large eddy simulation (LES) technique are discussed. The forced vibration test in a turbulent boundary layer is simulated. The models are force vibrated in the first antisymmetric modeto investigate the influences of a roof vibration on the wind pressure and flow field around a vibrating roof. The characteristics of unsteady aerodynamic forces in a wider range of reduced frequency of vibration are also investigated. A comparison between the wind tunnel experiment and the LES indicates that the LES can be used effectively to evaluate the unsteady aerodynamic force.

Wake Topology and Aerodynamic Performance of Heaving Wings

Simulation of the three-dimensional flow features generated by heaving wings is a great challenge for computational methods. The wake topology generated by oscillating rigid wings is studied. The unsteady laminar incompressible Navier-Stokes equations are solved on moving overlapping structured meshes using high-resolution numerical schemes. The numerical simulations are performed for a Reynolds number of Re=250 and different Strouhal numbers and heaving frequency.

Aeroelastic Stability of Turboprop Aircraft: Whirl Flutter

The whirl flutter, which is the specific type of dynamic aeroelastic stability phenomenon, is studied. Whirl flutter is caused by the effect of rotating parts of a turboprop power plant (propeller and gas turbine engine rotor). A brief historical overview and information regarding the occurrence of whirl flutter in the aerospace practice are presented. The physical principles of whirl flutter are explained on a simple mechanical system with two degrees of freedom. The analytical solution to determine the aerodynamic forces caused by the gyroscopic motion on each of the propeller blades is provided, and the influences of the main structural parameters on the whirl flutter stability are mentioned.

Goal- and Object-Oriented Models of the Aerodynamic Coefficients

The study introduces some new models for aerodynamic coefficients and demonstrates how the different models can be applied in object-oriented solutions. General aspects related to generation of the aerodynamic forces and moments are also covered. Modeling of linear and nonlinear, and steady and unsteady aerodynamic coefficient structures and forms, is discussed.

The Effects of Storage on Turbine Engine Fuels

Modern aviation requires reliable and safe sources of fuel. The variable composition of fuel depending upon crude oil source and refining methods creates challenges for fuel storage. Other problems are related to the presence of water, both from contamination through leaks in the storage tanks and the general solubility of water in jet fuel. The study addresses some problems associated with the incompatibility and long-term fuel storage.

12-Pulse Active Rectifier for More Electric Aircraft Applications

The study presents analysis and simulation of 12-pulse active rectifier for electric aircraft applications. Rather than using a fixed 400 Hz supply, variable frequency supply (360 to 800 Hz) is used depending on aircraft velocity.

The book promotes open discussion between research institutions, academia, and industry from around the globe on research and development of enabling technologies. The book covers many aspects of theory and simulation (physics, turbulence modeling, algorithms, applications, etc.) which deliver essential contributions, and provide their input and support to the cooperative efforts.

Dr. Konstantin Volkov
MEng, MSc, PhD, DSc, CEng, MIMechE, MInstP
Department of Mechanical and Automotive Engineering
School of Mechanical and Aerospace Engineering
Faculty of Science, Engineering and Computing
Kingston University, London, United Kingdom

Chapter 1

Learning from Nature: Unsteady Flow Physics in Bioinspired Flapping Flight

Haibo Dong, Ayodeji T. Bode-Oke and Chengyu Li

Additional information is available at the end of the chapter

http://dx.doi.org/10.5772/intechopen.73091

Abstract

There are few studies on wing flexibility and the associated aerodynamic performance of insect wings during free flight, which are potential candidates for developing bioinspired microaerial vehicles (MAVs). To this end, this chapter aims at understanding wing deformation and motions of insects through a combined experimental and computational approach. Two sets of techniques are currently being developed to make this integration possible: first, data acquisition through the use of high-speed photogrammetry and accurate data reconstruction to quantify the wing and body motions in free flight with great detail and second, direct numerical simulation (DNS) for force measurements and visualization of vortex structures. Unlike most previous studies that focus on the near-field vortex formation mechanisms of a single rigid flapping wing, this chapter presents freely flying insects with full-field vortex structures and associated unsteady aerodynamics at low Reynolds numbers. Our chapter is expected to lead to valuable insights into the underlying physics about flow mechanisms of low Reynolds number flight in nature, which will have great significance to flapping-wing MAV design and optimization research in the future.

Keywords: insect flight, high-speed photogrammetry, wing kinematics, wing flexibility, unsteady aerodynamics

1. Introduction

In nature, flying is a unique mechanism for generating control and maneuvering forces by flapping the wings. Weis-Fogh and Jensen [1] described flapping flight as a complex physical and biological problem that it is impossible to understand a single part of the process completely. One of the reasons is that the unsteady motion of wings has related flow mechanisms at a Reynolds number (*Re*) of 10 to 10^5 [2]. **Figure 1** illustrates a trend in the relationship between the Reynolds number and the body mass of both natural and man-made flying subjects. At this flow regime, lift producing mechanisms are intrinsically unsteady and vortex-dominated. Spanning over decades, considerable progress in understanding flapping flight has been achieved, and some general unsteady mechanisms have been identified. Examples include clap-and-fling [3–7], leading-edge vortex (LEV) [8–11], rotational lift [8, 12], wake capture [8, 13], wing-wing interactions, and body-wing interactions [14]. By using the rigid wing assumption, many of these mechanisms could explain the fluid phenomena near the wing, which are expressed as the motion of vortices, as well as the instantaneously local and resultant forces.

Flapping flight is a mode of transport widely adopted by natural fliers and has captured the interests of biologists and engineers because of several unique characteristics. From an energetic perspective, the propulsive efficiency of flapping motion can be higher than 85% [15]. Hence, flapping flight can be chosen as an alternative solution for aircraft propulsion to meet the need for high efficiency in energy consumption [16]. From the perspective of maneuverability and controllability, birds and insects have shown remarkable flying capabilities in tight spaces with multiple obstacles. In contrast, conventional aircraft cannot operate in such circumstances. The advantages of flapping flight have led to the development of micro air vehicles (MAVs) which mimic flapping flight. Many aspects of flapping flight, such as wing kinematics, structural response of wing, power consumption, and aerodynamics, are worth exploring. In particular, from a fluid dynamic point of view, the interaction of the flapping appendages with the surrounding air leads to the generation of vortices. Studying these vortex structures is of great importance for two reasons.

First, the vortex formation is related to the aerodynamic force, either indirectly or directly. Exerting forces on fluids can deform the fluids and leave footprints of the propulsion in the wake [17]. Therefore, even without force sensors and directly measuring pressure field, some elementary conclusions on force generation can be drawn based on the observed vortex formation in

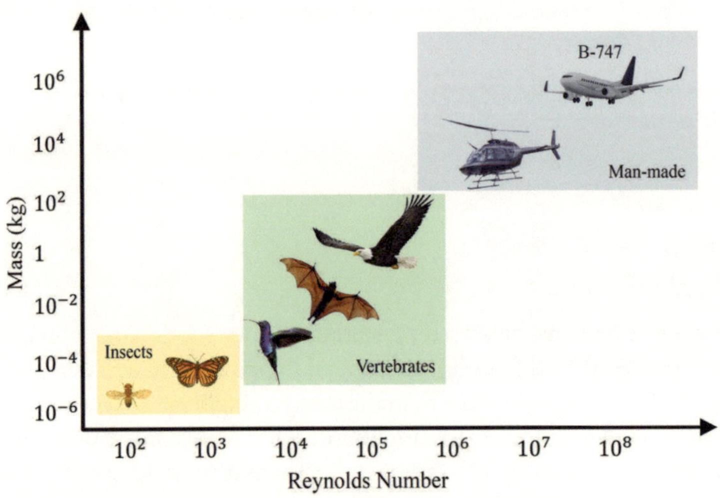

Figure 1. Characteristics of biological flapping flight and conventional man-made flight based on the Reynolds number and body mass. Picture courtesy of Science & Society Picture Library.

the wake. For example, it has been shown by [18] that the location and orientation of vortex streets behind an oscillating airfoil can indicate if the force acting on the airfoil is drag-indicative or thrust-indicative. When sufficient velocity and vorticity information around a submerged body is given, the wake survey methods can accurately predict force acting on the body [19–21]. Because force exerted by fluids on an immersed body is equal and opposite to the force exerted by the body to the fluids, controlling the vortex shedding in the wake of an airfoil is found to be effective in changing the aerodynamic performance of the airfoil [22].

Second, flapping flight usually operates in a low *Re* regime, and many of flow phenomena can be readily explained by using vortex dynamics. According to the well-known Biot-Savart law, vortex structures in far field can induce velocity change near a flapping wing without direct interaction. Tijdeman and Seebass [23] present an example of how vorticity in downstream induced velocity around an airfoil and caused the oscillation of lift lag behind the motion of the airfoil. A few but important near-field mechanisms for lift enhancement, that is, leading edge vortex [24], rotational forces, and wake capture [8], have also accounted for the presence of vortex structures and their interactions with flapping wings.

1.1. Insect wing and its motion

Insect wings are thin cuticular structures enforced by veins that spread across the wing in intricate patterns. The leading edge of the wing contains thickened veins that provide structural rigidity. These several radially stretched flexion lines on the wing represent regions of increased flexibility along which the wing can deform and yield variable camber [25]. Using a dragonfly forewing as an example, **Figure 2** shows the leading edge of the wing is enforced by multiple vein structures. Wing mass mostly arises from the wing venation, and the pattern of the wing venation varies among species with the wing to body mass ratio ranging between 0.5–4% in dipterans and hymenopterans and 3–10% in butterflies [26, 27].

The distribution of the wing mass has mechanical importance. Spanwise mass distribution defines the wing's moment of inertia about its hinge point and therefore indicates the power required for flapping the wings. The center of mass (CoM) of the insect wings usually lies at about 30–40% of the wing length from the wing hinge, reducing the wing moment of inertia. The chordwise distribution of the wing mass also is important in easing wing rotation at stroke reversal. The wing's center of mass is located below the longitudinal axis of rotation of the wing. Therefore, the inertial forces due to the wing acceleration help flip the wing at the stroke reversal.

Figure 2. Sketch of the main wing veins (dashed lines) on the left forewing.

The geometrical wing shape is also of great importance in the generation of the aerodynamic force. Wing total area directly affects the magnitude of the aerodynamic force. The wing loading, defined as the ratio of the body mass over the wing area, is an indicator of flight performance. Wing area tends to increase linearly with the body dimension, whereas body mass is a function of volume increasing with the cubed body size. Therefore, insects with larger bodies usually have higher wing loading. In addition to the area, the aspect ratio, defined as the ratio of the wingspan squared to the wing area, is used to describe the wing shape. The aspect ratio of the insect wings varies in a wide range from 2 for some butterflies, to 10 for some Odonata.

Insects modulate their wing kinematics to change the aerodynamic force magnitude and direction. The body motion also affects the net movement of the wing relative to the air and therefore influences the aerodynamics of flight. In maneuvering flights, for instance, the rotation of the body can cause significant asymmetry in the trajectory of the bilateral wings. This effect is more pronounced in low-flapping-frequency insects where the rotation of the body within one wing beat is significant.

1.2. Unsteady aerodynamic of flapping wing

To achieve efficient flights in a low Reynolds number regime, insects operate their wings with a combination of translational and rotational motion in a stroke plane. The dominant unsteady flow feature that is responsible for the aerodynamic force generation is the vortex formation close to the leading edge of flapping wings. This vortical structure is produced by a laminar flow separation and produces a region of low pressure on the wing toward the leading edge. Ellington et al. [24] first illustrated a direct evidence of the existence of this leading-edge vortex (LEV) by visualizing the flow around a three-dimensional robotic wing at Reynolds number around 10^3. This unique LEV is similar to the vortical structure produced during dynamic stall observed for conventional airfoils undergoing a rapid pitch motion. However, unlike the vortex produced during a dynamic stall, the LEV is not shed even after traveling many chord lengths of distance. As the flapping wing translates in its stroke plane, a spanwise velocity gradient interacts with the LEV. This causes the axial flow to spiral toward the wing tip direction. The axial flow transported momentum out of the vortex, keeping the LEV attached and stable. The LEV began to detach at the section close to the wing tip and shed into the wake. The vortex system generated by the flapping wing induces downwash in its surrounding fluid and forms a coherent momentum jet to maintain sustained flight.

1.3. Role of wing flexibility

Flying insects typically have flexible wings to adapt to the flow environment. **Figure 3** illustrates wing deformation of different species of insects with various wing geometry and aspect ratios under various flying modes. Due to the lack of internal musculature extending into the aerodynamic surface of the wing, insects have little active control over wing deformations. Therefore, the most surface morphology of insect wings is a product of the passive mechanical properties, while flapping wings interact with the inertial and aerodynamic forces. It is widely

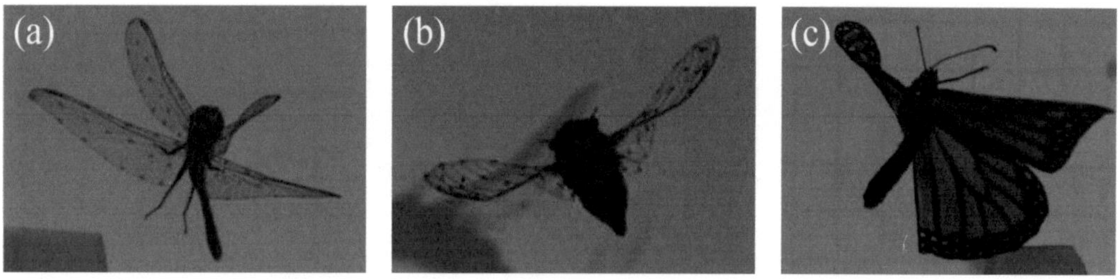

Figure 3. (a) Dragonfly in turning flight, (b) Cicada in forward flight, (c) Butterfly in takeoff flight; all showing large-scale wing deformation.

thought wing deformation would potentially provide new aerodynamic mechanisms of aerodynamic force productions over completely rigid wings in flying.

By applying either a two-dimensional foil or a highly idealized three-dimensional wing model [28–30], recent studies on the dynamic deformations during flapping flight mainly focused on the negative camber resulting from the aerodynamic and inertial forces. The development of high-speed photogrammetry has made the detailed measurements of wing deformation during high-frequency flapping motion possible. The study of deformable wing kinematics of locust [31] used a large number of marker points, and approximately 100 per wing shows that both forewings and hindwings were positively cambered on the downstroke through an "umbrella effect" whereby the trailing edge tension compressed the wing fan corrugated, reducing the projected area by 30% and releasing the tension in the trailing edge. The high-fidelity 3D dragonfly wing surface reconstruction performed by Koehler et al. [32] showed that insect wings could present up to 15% positive chordwise camber. Many fliers in nature have flexible wings which deform as the wings interact with the air around them. It has been opined that wing flexibility may provide new aerodynamic mechanisms of aerodynamic force production over completely rigid wings in flying [33–38].

1.4. Quantification of wing flexibility

The ability to capture the flight trajectory and flapping locomotion of flying insects is essential for studying flapping flight and quantifying the associated unsteady aerodynamics. Because most fliers flap too fast for the human eye to capture every detail, photogrammetry has been used to study birds [39], bats [40], and insects [41, 42].

Several previous studies have investigated the mesosurface morphological details of the wings of tethered [43, 44] and free-flying [45, 46] insects. However, these studies focused primarily on static wings. Laser scanning was used to measure the surface roughness of severed insect wings [47]. Dragonfly forewing and hindwing structures have been studied using a micro-CT scanner [48]. Corrugation in insect wings, for example, locust wings, has also been investigated [44]. However, to study corrugation, for a tethered locust, a large number of marker points (approximately 100 per hindwing) are used. Tracking these large numbers of marker points in free-flight studies is undesirable. Despite these quantitative visualizations of insect wing morphology, few works have been done on detailed measurements of 3D wing morphology during free flight. The small wing size, fast flapping motion, and unpredictability of insect movement complicate the tracking of the details of wing kinematics and deformation.

2. In vivo insect experiments and data acquisition

In this section, a unified methodology is introduced for the surface reconstruction of insect wings during free flying motion. The currently proposed method eliminates all rigid wing assumptions while minimizing the total number of tracking points in the outputted high-speed images from the photogrammetry system. The objective is to obtain a reconstructed insect modeled to capture the details of the real insect as much as possible, which will have implications in the flight aerodynamics.

2.1. High-speed videography

Image sequences of free-flying insects are collected using three synchronized Photron Fastcam SA3 60 K high-speed cameras capable of up to 1000 frames per second at a resolution of 1024 × 1024 pixels resolution with a shutter speed of 2 μs. Three cameras are fixed on an aluminum framework, as shown in **Figure 4a**. This setup allows us to reconstruct the insects' motion in a virtual space (**Figure 4b**).The framework and foundation can ensure that cameras are aligned orthogonal to each other on an optical breadboard. The slotted channels in framework allow us to adjust the distance between cameras and insects. For providing excellent temporal and spatial resolution, the cameras are positioned 1.5 m away from the dragonfly based on the body size and flapping frequency of the specimen. The optical breadboard not only allows us to mount our hardware to a sturdy anchor but also minimizes vibrations that will occur within the system. For the lighting system, two halogen photo optic lamps (OSRAM, 54,428) are chosen for our experiment.

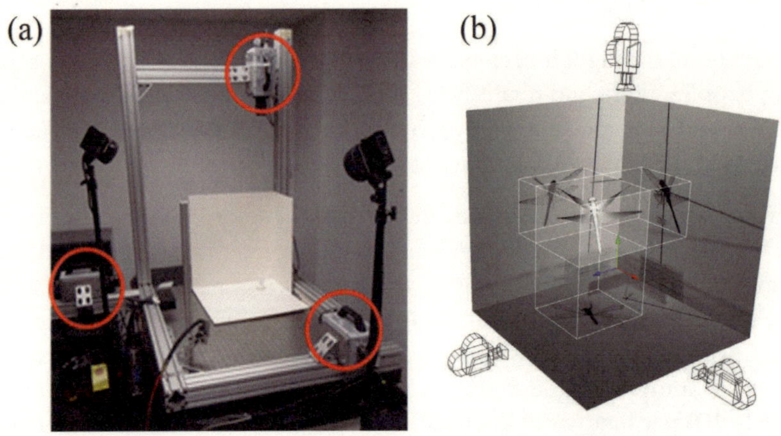

Figure 4. High-speed camera system setup in the laboratory (a) and in a virtual space (b).

After the camera system is installed, we use a daisy-chain method to loop the cameras to one another and trigger them by an external transistor-transistor logic (TTL) signal. This triggering method is an efficient way to minimize camera delay and maximize the response time. It allows us to capture images of a free-flying insect being synchronized and triggered simultaneously in three directions. The camera system can configure any standard personal computer, and the recorded sequences from the three cameras are stored locally on each camera's internal memory. All the data captured by the cameras are downloaded simultaneously to a desktop computer for subsequent analysis. The video sequences are named by using the number of cameras. Based on the quality of the image, the usable segments are identified for reconstruction.

The camera system was validated by evaluating the projection errors of filming geometries. The validation results show that there are less than 0.5 degrees in error over the entire filming data, and thus, the perspective errors can be ignored. Even though this system can minimize the human error associated with triggering the cameras, there are still some difficulties during the data collection. First, it is impossible to completely overcome the fact that the cameras are delayed from frame to frame by a small variation. Moreover, wing surface reflections caused by the lighting system sometimes make it difficult to identify the marker point clearly in the images. Since the flight paths of insects are unpredictable, capturing the true voluntary flight motion in a certain focus range of cameras is the greatest challenge.

2.2. 3D surface reconstruction

For reconstructing the wing kinematics and deformation, each insect wing is marked with a fine-tipped permanent marker before shooting the videos (e.g., a dragonfly as shown in **Figure 5a**). Since the added weight of the ink on the surface of the wing is small, we assume it is negligible and does not affect the flight performance. For an arbitrary point on a dragonfly's wing in each frame, we use the perspective projection method to decide its location in multiple projection planes. The photogrammetry system is used to capture the insect in flight.

The initial 3D wing template models are generated with Catmull-Clark subdivision surfaces by using a computer graphics software Autodesk Maya (as shown in **Figure 5b**). Based on the high-speed films, we align first-level vertices of the subdivision surface hierarchy corresponding to the marker points on insects' wings (e.g., dragonfly's forewings and hindwings). After the initial template surfaces of wings are generated, they are recorded as a keyframe animation. By repeatedly adjusting the anchor point-based alignment process along with each axis for each time step, the first level vertices of wings are completed. Although the whole process of wing reconstruction is a bit labor intensive, it is currently the only effective way to reconstruct a deformable, quad-winged insect in free flight. **Figure 6** presents the front and side views of the reconstructed wings overlapping with the corresponding high-speed images. Thus, the approximation of the 3D wing shapes such as spanwise bend, chordwise bend, and twist can be captured with smooth subdivision surface representation. Comparing with tethered insects, free-flying insects present many challenges to the surface reconstruction work due to the nonlinear translation and rotation motion, especially during turning maneuvers. **Figure 7** visualizes a reconstructed motion of the dragonfly during the free-flight maneuver at selected instants.

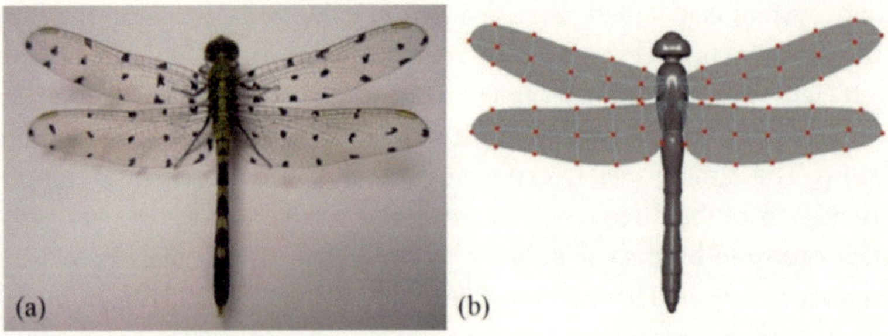

Figure 5. Initial configuration of a dragonfly template mesh. (a) Dragonfly with marker points on its wings. (b) Wing and body template models [49].

2.3. Numerical method

To study the aerodynamics of free-flight insects, the flow fields were generated by direct numerical simulations of the three-dimensional unsteady, viscous incompressible Navier-Stokes equations, as written in the following equations.

$$\frac{\partial u_i}{\partial x_i} = 0 \tag{1}$$

$$\frac{\partial u_i}{\partial t} + \frac{\partial (u_i u_j)}{\partial x_j} = -\frac{\partial p}{\partial x_i} + \frac{1}{Re}\frac{\partial}{\partial x_j}\left(\frac{\partial u_i}{\partial x_j}\right) \tag{2}$$

where u_i $(i = 1, 2, 3)$ are the velocity components in the x-, y-, and z-directions, respectively, p is the pressure, and Re is the Reynolds number.

The above Navier-Strokes equations are discretized using a cell-centered, collocated (non-staggered) arrangement, where the velocity components and pressure are located at the same physical location. The equations are then solved by using the fractional step method. The discretization of the convective terms and diffusion terms are achieved by using an Adams-Bashforth scheme and an implicit Crank-Nicolson scheme, respectively. The immersed boundary method is a computational method used to simulate fluid flow over bodies which are embedded within a Cartesian grid. It eliminates the need for the complicated re-meshing

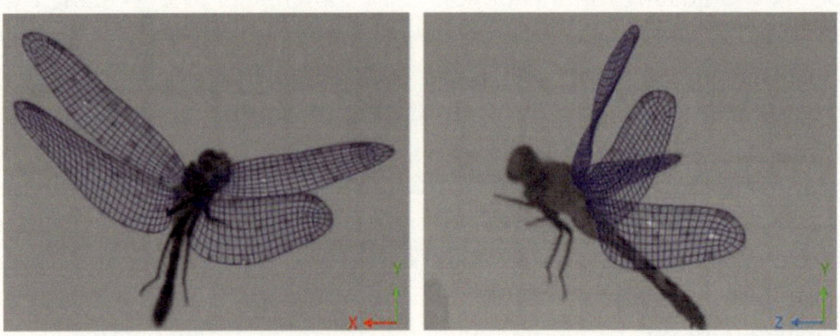

Figure 6. Reconstructed wings at a time step where a large amount of twist and camber is present in multiple wings.

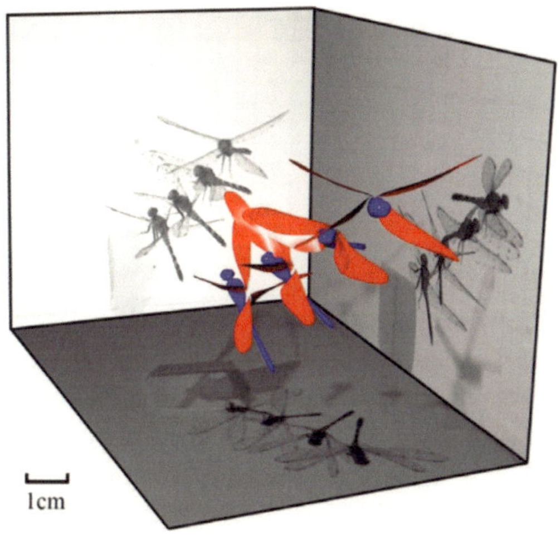

Figure 7. Motion reconstruction of dragonfly turning maneuver. The side panels show 4 of 116 frames recorded by high-speed videography [49].

algorithms and reduces the computational cost for the mesh generation in each time step that is usually employed by conventional body-conformal methods. More details for this current numerical approach can be found in [50]. The current in-house solver has also been validated by simulating canonical revolving/flapping plates [51–56], the flapping wings of insects [14, 49, 57, 58], and physiological flows [59].

3. High-fidelity analysis of flapping flight aerodynamics

In previous sections, we presented the data acquisition methods developed to obtain the most realistic reconstruction from high-speed videos. The output is a high-fidelity 3D model with the wing and body motions encoded therein. In the following discussion, the motion and deformation metrics are not isolated to study their effect on flight performance; rather, the deformations are intrinsic and influence aerodynamic footprint of the insect. Naturally, the next step is to simulate the flapping locomotion and identify the associated wake structures. We use computational fluid dynamics (CFD) simulation to understand the relevant flow features of different insects in free flight.

3.1. Dragonfly in takeoff turning flight

Dragonflies are aerial predators and feed on other flying insects. Unlike most other insects, such as flies, wasps, and cicadas, that have either reduced hindwings or functionally combined forewings and hindwings as a single pair, dragonflies have maintained two pairs of wings throughout their evolution [60]. Their neuromuscular systems allow them to individually change many aspects of wing motion in each single wing, including the angle of attack, stroke amplitude, and wing deviation, which gives them unique flying capabilities of flight control.

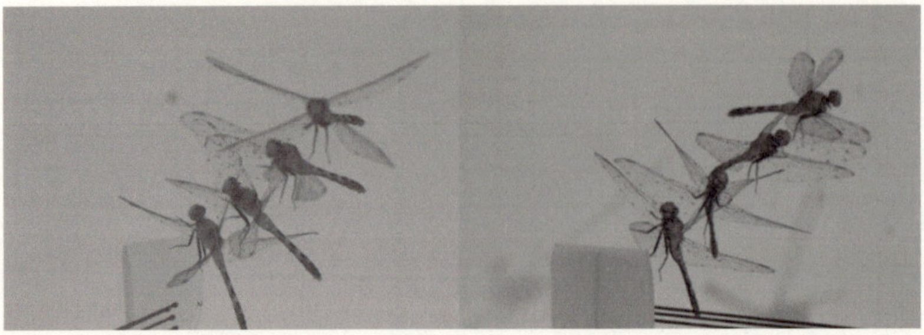

Figure 8. Snapshots of a dragonfly in takeoff turning flight from the front-view camera (left) and side-view camera (right).

In general, the flapping motion shown in **Figure 8** generates pronounced changes in the angle of attack between each side wings, especially for the forewings. During the downstroke, the magnitude of the angle of attack for the left forewing is $29 \pm 3^{\circ}$, whereas the value for the right forewing is $43 \pm 5^{\circ}$. During the upstroke, the variations of the angle of attack for the left and right forewings are $73 \pm 5^{\circ}$ and $49 \pm 3^{\circ}$, respectively. The angle of attack of the left and right hindwings at the mid-downstroke is $36 \pm 1^{\circ}$ and $21 \pm 3^{\circ}$, respectively. During the upstroke, the minimum value for the left hindwing is $23 \pm 4^{\circ}$ and for the right hindwing is $33 \pm 4^{\circ}$. For both forewings and hindwings, the asymmetric wing motion results in a relatively large angle of attack on the left-side wings during the downstroke and a small one during the upstroke. This finding implies that compared to the right-side wings, the left-side wings might experience higher drag force during the downstroke and lower thrust force during the upstroke.

Figure 9 shows the time sequence of the 3D flow field, which is identified by plotting the iso-surface of the Q-criterion [61]. To illustrate the development of the vortical structures, six snapshots from the flapping motion are shown. For each wing, a leading-edge vortex (LEV) is developed and grows stronger, remaining stably attached to the wing during the downstroke. As the wing sweeps, the LEV, the tip vortex (TV), and the shed trailing-edge vortex (TEV) connect and form a vortex loop. Because of the phase relationship between the forewings and hindwings, when the forewings reach the end of their downstroke, the hindwings have already started to move upward. As the hindwings flap upward, distinct fully developed vortex rings are gradually shed into the flow field from the trailing edge of the wings. At the same

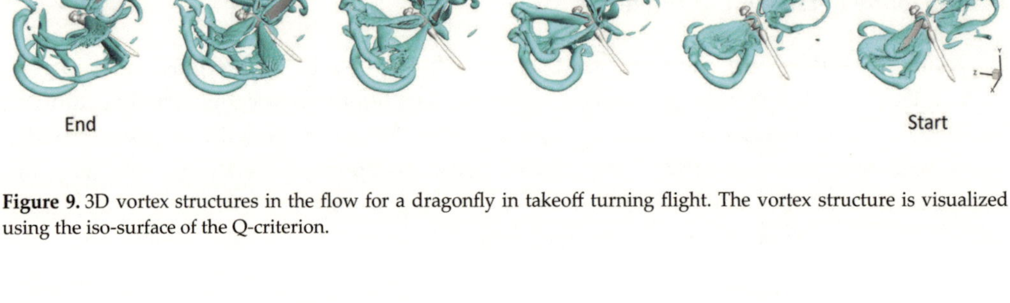

Figure 9. 3D vortex structures in the flow for a dragonfly in takeoff turning flight. The vortex structure is visualized using the iso-surface of the Q-criterion.

time, the upward-moving hindwings interact with the vortex loop formed by the forewings. This flow feature has been termed forewing-hindwing interaction in previous 2D and 3D tandem-wing studies [38, 62]. In addition to the forewing-hindwing interaction, during the upstroke, the wings catch their own wakes from the preceding downstroke, which disrupts the vortex loop structures through the wing-wake interaction and forms a stronger TV and TEV. During the maneuver, distinct asymmetric vortex formation also occurs between the left and right sides. This asymmetric phenomenon also makes the shed vortex rings tilted and distorted. By interacting with the vortex loops formed by other wings as well as previously shed vortex loops, the wake becomes more complicated. Due to the viscous dissipation effect and wing-wake interactions, only the LEV and TV in the near wake are still distinguishable in the flow field. The key features observed here are the presence of vortex loop structures in the near wake around the wings.

3.2. Butterfly in vertical takeoff

The flapping motion of a monarch butterfly (*Danaus plexippus*) in vertical takeoff flight is present in **Figure 10**. The butterfly's body and wing were then reconstructed with extraordinary details. Direct numerical simulation was then carried out in order to understand the vortex formation during the takeoff motion.

The vortex structures of the flow field are shown in **Figure 11**. Several thin swirling vortices start from each wing tip as the wings flap downward. The thin swirling vortices twist immediately after the separation from each wing tip and form a tip vortex (TV) during the downstroke. As the butterfly left upward, the thin vortices are merged by the viscosity into a coherent vortex under the insect body. The TV during downstroke generates lift as the reaction of inducing the downward flow. Wingtip vortices during upstroke are also made by the vorticities of the wings aligned close to the negative and positive vertical directions. Trailing-edge vortices released at the transitions from downstroke to upstrokes are barely visible in the wake of butterfly flapping motion.

Figure 10. Snapshots of a butterfly in takeoff flight from the front-view camera (left) and side-view camera (right).

Figure 11. 3D vortex structures in the flow for a butterfly in takeoff flight. The vortex structure is visualized using the iso-surface of the Q-criterion.

3.3. Damselfly in yaw turn

Here, a damselfly (*Calopteryx maculata*) is involved in performing a yaw turn maneuver during which it also ascends for about three body lengths (**Figure 12**). To perform a turning maneuver, the wings must generate aerodynamic forces to sustain body weight while simultaneously producing turning moments to rotate the body around its center of mass (CoM) while remaining almost-stationary or moving forward. During a yaw turn, the horizontal component of the aerodynamic force is oriented toward the center of curvature. This force reorientation produces lateral forces for rotation around the center of mass. In addition, an asymmetry in wing kinematics between the wings on the outside of the turn (left wings) and right wings on the inside of the turn is necessary to create a yaw torque differential. Insights into how the damselfly flight forces can be gleaned from the flow field data. To generate forces for flight, the damselfly used an unsteady mechanism such as a leading-edge vortex on its wings which feeds into a tip vortex (**Figure 13**). Although the wake structure is quite complex, an asymmetric flow structure with a stronger flow field oriented toward the right wing is observed, which indicates that the inner wings may be playing a substantial role in executing the turn by creating large force/yaw torque differences between the contralateral wing pairs.

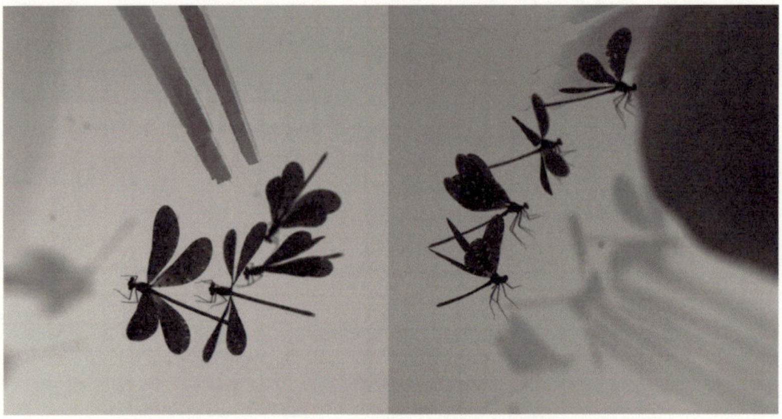

Figure 12. Snapshots of a damselfly in yaw turn from the top-view camera (left) and side-view camera (right).

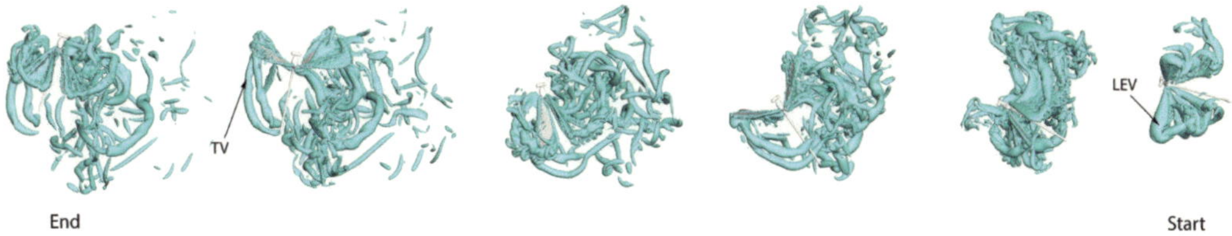

Figure 13. 3D vortex structures in the flow for a damselfly in yaw turn. The vortex structure is visualized using the iso-surface of the Q-criterion.

3.4. Cicada in turning maneuver

Here, the flight of a cicada which initiates flight and immediately executes a banked turn is recorded. Unlike the yaw turn scenario of the damselfly where the wings do the majority of the force reorientation, and the roll motion of the body is minimal, banked turns are distinct but more commonly found in nature. During banked turns, the animal reorients the total aerodynamic force vector toward the center of the turn by simply rotating the body around its longitudinal axis. This is evident in **Figure 14** wherein the cicada rolls its body by about 90° to reorient the force vector into the center of curvature like an airplane. During this flight, the cicada still utilizes unsteady mechanism such as a LEV (**Figure 15**) to help maintain a strong enough component to sustain weight. The asymmetry in flow features on the left and right sides indicates a force difference necessary to induce a roll torque for the turn.

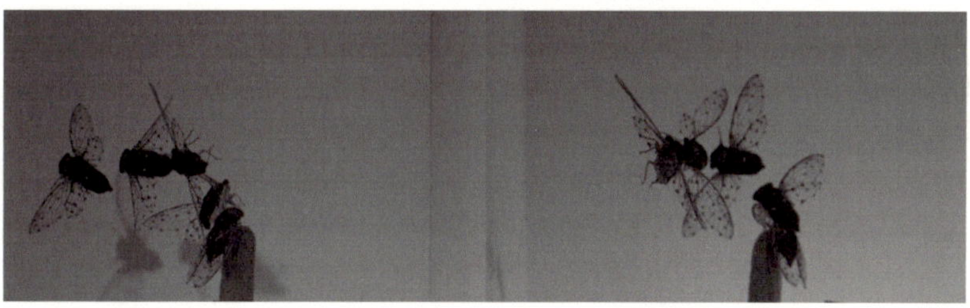

Figure 14. Snapshots of a cicada in takeoff turning flight from the front-view camera (left) and side-view camera (right).

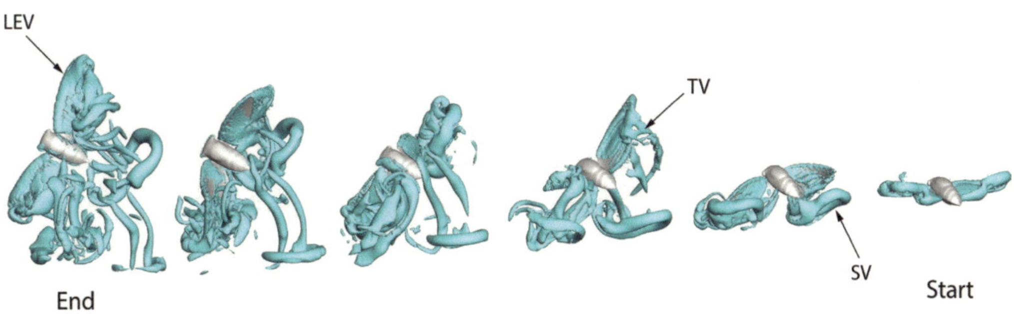

Figure 15. 3D vortex structures in the flow for a cicada in takeoff turning flight. The vortex structure is visualized using the iso-surface of the Q-criterion.

4. Conclusion

In the current chapter, we present an integrative approach to study the aerodynamics of free-flying insects. This proposed novel approach combines high-speed photogrammetry, 3D surface reconstruction, and computational fluid dynamics and is capable of measuring the wing kinematics and surface deformation and simulating its aerodynamic performance. Four different species of insects, including dragonfly, damselfly, butterfly, and cicada, were investigated to demonstrate the capability of the current method. From these collected data, we provide insight into the understanding of flapping wing mechanism and revealing the underlying flow physics for future MAV designs and applications.

Acknowledgements

This research is supported by NSF grant CBET-1313217 and AFOSR grant FA9550-12-1-0071. We thank former members of Flow Simulation Research Group at Wright State University and University of Virginia for capture of highspeed videos. We thank Yan Ren for **Figure 4(b)**.

Author details

Haibo Dong*, Ayodeji T. Bode-Oke and Chengyu Li

*Address all correspondence to: haibo.dong@virginia.edu

Mechanical and Aerospace Engineering, University of Virginia, Charlottesville, VA, USA

References

[1] Weis-Fogh T, Jensen M. Biology and physics of locust flight. i. Basic principles in insect flight. A critical review. Philosophical Transactions of the Royal Society of London Series B: Biological Sciences. 1956;**239**:415-458

[2] Dudley R. The Biomechanics of Insect Flight. Princeton: Princeton University Press; 2000

[3] Weisfogh T. Quick estimates of flight fitness in hovering animals, including novel mechanisms for lift production. Journal of Experimental Biology. 1973;**59**:169-230

[4] Lighthill MJ. On the Weis–Fogh mechanism of lift generation. Journal of Fluid Mechanics. 1973;**60**:1-17

[5] Ellington CP. The aerodynamics of hovering insect flight. 3. Kinematics. Philosophical Transactions of the Royal Society of London Series B: Biological Sciences. 1984;**305**:41-78

[6] Maxworthy T. Experiments on the Weis–Fogh mechanism of lift generation by insects in hovering flight Part 1. Dynamics of the 'fling'. Journal of Fluid Mechanics. 1979;**93**:47-63

[7] Ellington CP. Unsteady aerodynamics of insect flight. Symposia of the Society for Experimental Biology. 1995;**49**:109-129

[8] Dickinson MH, Lehmann FO, Sane SP. Wing rotation and the aerodynamic basis of insect flight. Science. 1999;**284**:1954-1960

[9] Birch. Investigation of the near-field tip vortex behind an oscillating wing. Journal of Fluid Mechanics 2005;**544**:201-241

[10] Maxworthy T. The fluid-dynamics of insect flight. Annual Review of Fluid Mechanics. 1981;**13**:329-350

[11] van den Berg C, Ellington CP. The vortex wake of a 'hovering' model hawkmoth. Philosophical Transactions of the Royal Society of London Series B: Biological Sciences 1997;**352**:317-328

[12] Sane SP, Dickinson MH. The aerodynamic effects of wing rotation and a revised quasi-steady model of flapping flight. The Journal of Experimental Biology. 2002;**205**:1087-1096

[13] Sun M, Tang J. Unsteady aerodynamic force generation by a model fruit fly wing in flapping motion. The Journal of Experimental Biology. 2002;**205**:55-70

[14] Liu G, Dong H, Li C. Vortex dynamics and new lift enhancement mechanism of wing–body interaction in insect forward flight. Journal of Fluid Mechanics. 2016;**795**:634-651

[15] Anderson JM, Streitlien K, Barrett DS, Triantafyllou MS. Oscillating foils of high propulsive efficiency. Journal of Fluid Mechanics. 1998;**360**:41-72

[16] Shyy W, Berg M, Ljungqvist D. Flapping and flexible wings for biological and micro air vehicles. Progress in Aerospace Sciences. 1999;**35**:455-505

[17] Platzer MF, Jones KD, Young J, Lai JCS. Flapping-wing aerodynamics: Progress and challenges. AIAA Journal. 2008;**46**:2136-2149

[18] Jones K, Dohring C, Platzer M. Wake structures behind plunging airfoils: A comparison of numerical and experimental results. AIAA Paper 1996-0078:1996

[19] Wu JZ. Theory for aerodynamic force and moment in viscous flows. AIAA Journal. 1981;**19**:432-441

[20] Noca F, Shiels D, Jeon D. A comparison of methods for evaluating time-dependent fluid dynamic forces on bodies, using only velocity fields and their derivatives. Journal of Fluids and Structures. 1999;**13**:551-578

[21] Wu JZ, Ma HY, Zhou MD. Vorticity and Vortex Dynamics. Berlin Heidelberg New York: Springer; 2006

[22] Ho S, Nassef H, Pornsinsirirak N, Tai YC, Ho CM. Unsteady aerodynamics and flow control for flapping wing flyers. Progress in Aerospace Sciences. 2003;**39**:635-681

[23] Tijdeman H, Seebass R. Transonic flow past oscillating airfoils. Annual Review of Fluid Mechanics. 1980;**12**:181-222

[24] Ellington CP, van den Berg C, Willmott AP, Thomas ALR. Leading-edge vortices in insect flight. Nature 1996;**384**:626-630.

[25] Dudley R. The Biomechanics of Insect Flight: Form, Function, Evolution. Princeton: Princeton University Press; 2002

[26] Ellington CP. The aerodynamics of hovering insect flight. 1. The quasi-steady analysis. Philosophical Transactions of the Royal Society of London Series B: Biological Sciences. 1984;**305**:1-15

[27] Betts C, Wootton R. Wing shape and flight behaviour in butterflies (Lepidoptera: Papilionoidea and Hesperioidea): A preliminary analysis. The Journal of Experimental Biology. 1988;**138**:271-288

[28] Chimakurthi SK, Tang J, Palacios R, Cesnik CES, Shyy W. Computational aeroelasticity framework for analyzing flapping wing micro air vehicles. AIAA Journal. 2009;**47**:1865-1878

[29] Hamamoto M, Ohta Y, Hara K, Hisada T. Application of fluid-structure interaction analysis to flapping flight of insects with deformable wings. Advanced Robotics. 2007;**21**:1-21

[30] Yin B, Luo HX. Effect of wing inertia on hovering performance of flexible flapping wings. Physics of Fluids. 2010;**22**:111902

[31] Walker SM, Thomas ALR, Taylor GK. Deformable wing kinematics in the desert locust: How and why do camber, twist and topography vary through the stroke? Journal of the Royal Society Interface. 2009;**6**:735-747

[32] Koehler C, Liang ZX, Gaston Z, Wan H, Dong HB. 3D reconstruction and analysis of wing deformation in free-flying dragonflies. The Journal of Experimental Biology. 2012;**215**:3018-3027

[33] Wootton RJ. Leading edge section and asymmetric twisting in the wings of flying butterflies (Insecta, Papilionoidea). The Journal of Experimental Biology. 1993;**180**(1):105-117

[34] Combes SA, Daniel TL. Shape, flapping and flexion: wing and fin design for forward flight. Journal of Experimental Biology. 2001;**204**(12):2073-2085

[35] Combes SA, Daniel TL. Flexural stiffness in insect wings. I. Scaling and influence of wing venation. Journal of Experimental Biology. 2003;**206**:2979-2987

[36] Combes SA, Daniel TL. Flexural stiffness in insect wings. II. Spatial distribution and dynamic wing bending. Journal of Experimental Biology. 2003;**206**:2989-2997

[37] Sane SP. The aerodynamics of insect flight. Journal of Experimental Biology 2003;**206**(23): 4191-4208

[38] Lehmann FO. When wings touch wakes: Understanding locomotor force control by wake-wing interference in insect wings. The Journal of Experimental Biology. 2008;**211**:224-233

[39] Tobalske BW, Warrick DR, Clark CJ, Powers DR, Hedrick TL, Hyder GA, et al. Three-dimensional kinematics of hummingbird flight. The Journal of Experimental Biology. 2007;**210**:2368-2382

[40] Norberg UML, Winter Y. Wing beat kinematics of a nectar-feeding bat, *Glossophaga soricina*, flying at different flight speeds and Strouhal numbers. The Journal of Experimental Biology. 2006;**209**:3887-3897

[41] Ristroph L, Berman GJ, Bergou AJ, Wang ZJ, Cohen I. Automated hull reconstruction motion tracking (HRMT) applied to sideways maneuvers of free-flying insects. The Journal of Experimental Biology. 2009;**212**:1324-1335

[42] Liu YP, Sun M. Wing kinematics measurement and aerodynamics of hovering droneflies. The Journal of Experimental Biology. 2008;**211**:2014-2025

[43] Sunada S, Song D, Meng X, Wang H, Zeng L, Kawachi K. Optical measurement of the deformation, motion, and generated force of the wings of a moth. JSEM International Journal. 2002;**45**(4):836-842

[44] Walker SM, Thomas ALR, Taylor GK. Photogrammetric reconstruction of high-resolution surface topographies and deformable wing kinematics of tethered locusts and free-flying hoverflies. Journal of the Royal Society Interface. 2009;**6**:351-366

[45] Bergou AJ, Ristroph L, Guckenheimer J, Cohen I, Wang ZJ. Fruit flies modulate passive wing pitching to generate in-flight turns. Physical Review Letters. 2010;**104**:141801

[46] Walker SM, Thomas ALR, Taylor GK. Deformable wing kinematics in free-flying hoverflies. Journal of the Royal Society Interface. 2010;**7**:131-142

[47] Sudo S, Tsuyuki K, Kanno K. Wing characteristics and flapping behavior of flying insects. Experimental Mechanics. 2005;**45**:550-555

[48] Jongerius SR, Lentink D. Structural analysis of a dragonfly wing. Experimental Mechanics. 2010;**50**:1323-1334

[49] Li C, Dong H. Wing kinematics measurement and aerodynamics of a dragonfly in turning flight. Bioinspiration & Biomimetics. 2017;**12**:026001

[50] Mittal R, Dong H, Bozkurttas M, Najjar FM, Vargas A, von Loebbecke A. A versatile sharp interface immersed boundary method for incompressible flows with complex boundaries. Journal of Computational Physics 2008;**227**:4825-4852.

[51] Li C, Dong H. Three-dimensional wake topology and propulsive performance of low-aspect-ratio pitching-rolling plates. Physics of Fluids. 2016;**28**:071901

[52] Xu M, Wei M, Li C, Dong H. Adjoint-based optimization of flapping plates hinged with a trailing-edge flap. Theoretical and Applied Mechanics Letters. 2015;**5**:1-4

[53] Wan H, Dong H, Li C, Liang Z. Vortex formation and aerodynamic force of low aspect-ratio plate in translation and rotation. AIAA Paper 2012-3278; 2012

[54] Li C, Dong H. Wake structure and aerodynamic performance of low aspect-ratio revolving plates at low Reynolds number. AIAA Paper 2014-1453; 2014

[55] Li C, Dong H, Liang Z. Proper orthogonal decomposition analysis of 3-D wake structures in a pitching-rolling plate. AIAA Paper 2016-2071; 2016

[56] Li C, Dong H, Cheng B. Effects of aspect ratio and angle of attack on tip vortex structures and aerodynamic performance for rotating flat plates. AIAA Paper 2017-3645; 2017

[57] Li C, Wang J, Dong H. Proper orthogonal decomposition analysis of flapping hovering wings. AIAA Paper 2017-0327; 2017

[58] Wang J, Li C, Ren Y, Dong H. Effects of surface morphing on the wake structure and performance of flapping plates. AIAA 2017-3643; 2017

[59] Li C, Jiang J, Dong H, Zhao K. Computational modeling and validation of human nasal airflow under various breathing conditions. Journal of Biomechanics. 2017;**64**:59-68

[60] Bomphrey RJ, Nakata T, Henningsson P, Lin H-T. Flight of the dragonflies and damselflies. Philosophical Transactions of the Royal Society of London Series B: Biological Sciences. 2016;**371**:20150389

[61] Hunt JCR, Wray AA, Moin P. Eddies, streams, and convergence zones in turbulent flows. Center for Turbulence Research Report CTR-S88; 1988:193-208

[62] Wang J, Russell D. Effect of forewing and hindwing interactions on aerodynamic forces and power in hovering dragonfly flight. Physical Review Letters. 2007;**99**:148101

Chapter 2

Helicopter Flight Physics

Constantin Rotaru and Michael Todorov

Additional information is available at the end of the chapter

http://dx.doi.org/10.5772/intechopen.71516

Abstract

This chapter is dedicated to present the principles that constitute the fundamentals of helicopter flight physics, starting from the basics of the main rotor aerodynamics and of the component parts related to flight control. The chapter opens with a short history of helicopter development, taking the date of 13th November 1907 for a reference point; this is the date when the first helicopter flight occurred, having the French man, Paul Cornu, for a pilot. The main constructive solutions for helicopters are presented and the basic equations of fluid mechanics are applied on a helicopter model with one main rotor and tail rotor. Helicopter hovering, vertical flight, and forward flight are approached, too, one by one. Furthermore, the ground effect, autorotation, stability, and helicopter control are focused on. At the end of the chapter, the main factors that determine the helicopter performances are mentioned.

Keywords: helicopter aerodynamics, induced velocity, autorotation, ground effect, hover

1. Introduction

The helicopter belongs to the flight machine category with the highest operational efficiency because it does not need special take-off and landing grounds with expensive utilities and logistics equipment. For the short and medium range, the flight efficiency of helicopters is comparable with those of the airplanes. It is able to hover, fly sideward, backward, forward, and perform other desirable maneuvers in civilian field like sea and mountain rescue, police surveillance, and firefighting; or in military missions such as battlefield surveillance, troop transport, assault, and antitank operations. So far with the help of helicopters, lives of over a million of people were saved. In the last years, the results obtained in the scientific research of many aeronautical disciplines has allowed for large increase in the flight dynamics, control, navigation, and lift capabilities of helicopters.

The aerodynamic limitations imposed by the main rotor were understood better and overcome gradually so, the present helicopters are able to fly at about 370 km/h. The continued advance in the computer-aided design, manufacturing, and lightweight materials have permitted new approaches in the helicopter configuration concepts and design. The helicopter lift force is provided by the main rotor with the blades that spin about the shaft and all the flight maneuvers under the pilot's full control suppose a significant mechanical and aerodynamic complexity.

The word "helicopter" comes from two Greek words, "helliko" (spiral) and "pteron" (wing). The idea of vertical flight could be localized in time, in the years of about 400 BC, when was built so called "Chinese top," consisted of feathers at the end of a stick which was spun between the hands to generate lift. In 1483, Leonardo da Vinci proposed a flight device, which comprised a helical surface formed out of iron wire. According to the historical sources, in about 1754, Mikhail Lomonosov of Russia had built a coaxial rotor, modeled after the Chinese top, but powered by a spring device, which flew freely.

A short list of the most important achievements in the historical evolution of helicopters is the following:

1843: Sir George Cayley (considered the inventor of the airplane) published a paper, where he gives some scientific details about the vertical flight of the aircraft;

1860: Ponton d"Amecourt of France built a number of small steam-powered helicopter models;

1874: Wilheim von Achenbach of Germany built a single rotor model and he had the idea to create a sideward thrusting tail rotor in order to counteract the main rotor torque reaction;

1880: Thomas Alva Edison tested several rotor configurations powered by an electric motor;

Four years after Orville Wright first successful powered flight, which took place in December 17, 1903, a French, named Paul Cornu constructed a helicopter and *flew for the first time in the world in November 13, 1907*;

1907: the French brothers Louis and Jaques Breguet built a helicopter (quad rotor, in the form of a horizontal cross) powered by a 40-hp. engine. This helicopter did not fly completely free due to its lack of stability;

1909: Igor Ivanovitch Sikorsky built a nonpiloted coaxial helicopter prototype;

1912: Boris Yuriev tried to build a helicopter with a single main rotor and tail rotor configuration. He proposed the concept of cyclic pitch for rotor control;

1914: the Danish Jen C. Ellehammer designed a helicopter with coaxial rotors. The aircraft made several short hops but never made a properly flight;

1917: Stephan Petroczy (Austrian) build and flew a coaxial rotor helicopter;

1919: Henry Berliner (USA) built a counter-rotating coaxial helicopter;

1920: Raul Pescara (Argentina) built a coaxial helicopter;

1922: Georges des Bothezat (USA) designed and built a helicopter for the USA army. He was the first specialist who described the helicopter autorotation;

http://dx.doi.org/10.5772/intechopen.71516

1939: Igor Ivanovitch Sikorsky built the helicopter VS-300 which flew in May 13, 1940. He could be considered the most important person in the helicopter design.

1.1. Helicopter configurations

The helicopter is a complex aircraft that obtains both lift and thrust from blades rotating about a vertical axis. The term "rotary wing" is often used to distinguish the helicopter from airplane, which is a "fixed wing" aircraft. The helicopter can have one or more engines, and it uses gear boxes connected to the engines by rotating shafts to transfer the power from engines to the rotors (**Figure 1**).

The most common helicopter configuration consists of one main rotor as well as a tail rotor to the rear of the fuselage (**Figure 2a**). A tandem rotor helicopter has two main rotors; one at the front of the fuselage and one at the back (**Figure 2b**). This type of configuration does not need a tail rotor because the main rotors are counter rotating. It was proposed by the Serbian man Dragoljub Ivanovich in 1953.

A variant of the tandem is the coaxial rotor helicopter (**Figure 3a**) which has the same principle of operation, but the two main rotors are mounted one above the other on coaxial rotor shafts. This constructive solution was developed by Nicolai Ilich Kamov. Another helicopter type is the synchropter, which use intermeshing blades (**Figure 3b**). This type of helicopter was proposed by Charles Kaman.

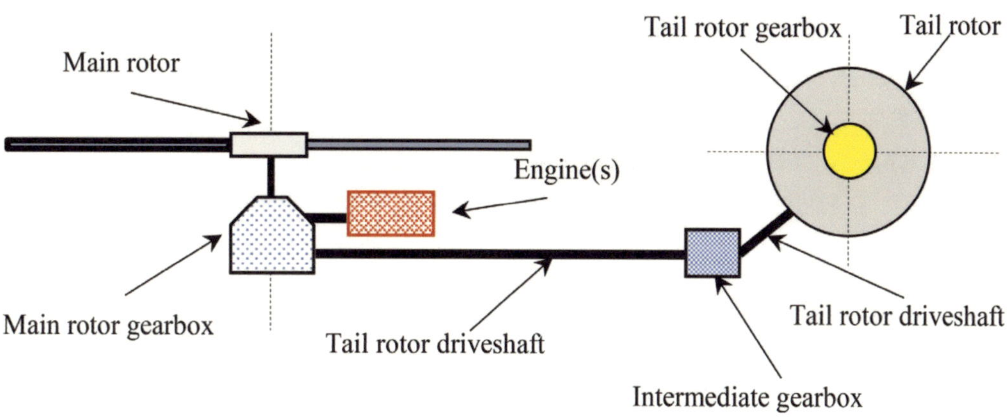

Figure 1. Typical helicopter drive train.

Figure 2. The single main rotor (a) and the tandem rotor helicopter (b).

Figure 3. The coaxial rotors (a) and the intermeshing blades (b).

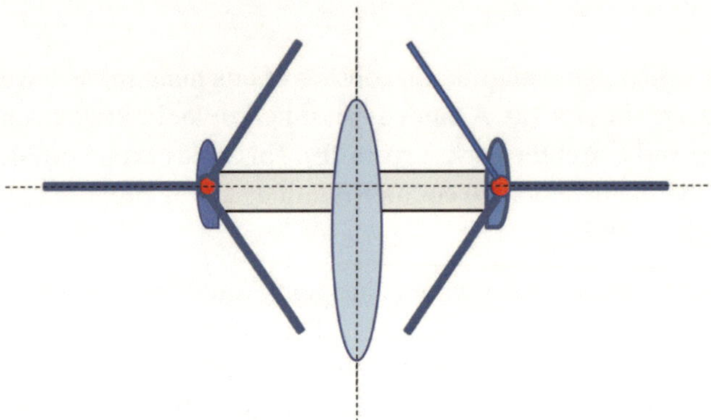

Figure 4. The side by side rotors.

If the two rotors are mounted either side of the fuselage, on pylons or wing tips, the configuration is referred to as side by side (**Figure 4**).

Another aircraft type that should be mentioned is the autogiro (invented by Huan de la Cievra), which is a hybrid between a helicopter and a fixed wing airplane. It uses a propeller for the forward propulsion and has freely spinning nonpowered main rotor that provides lift.

2. Basics of helicopter aerodynamics

The basic flight regimes of helicopter include hover, climb, descent, and forward flight, and the analysis and study of these flight regimes can be approached by the actuator disk theory, where an infinite number of zero thickness blades support the thrust force generated by the rotation of the blades [1]. The air is assumed to be incompressible and the flow remains in the same direction (one-dimensional), which for most flight conditions is appropriate. The helicopter main rotor generates a vertical force in opposition to the helicopter's weight and a horizontal propulsive force for forward flight. Also, the main and tail rotors generate the forces and moments to control the attitude and position of the helicopter in three-dimensional space.

2.1. Hovering flight

The cross sections in **Figure 5** denote: the plane far upstream of the rotor, where in the hovering case the air velocity is null (section 0–0); the planes just above and below the rotor

http://dx.doi.org/10.5772/intechopen.71516

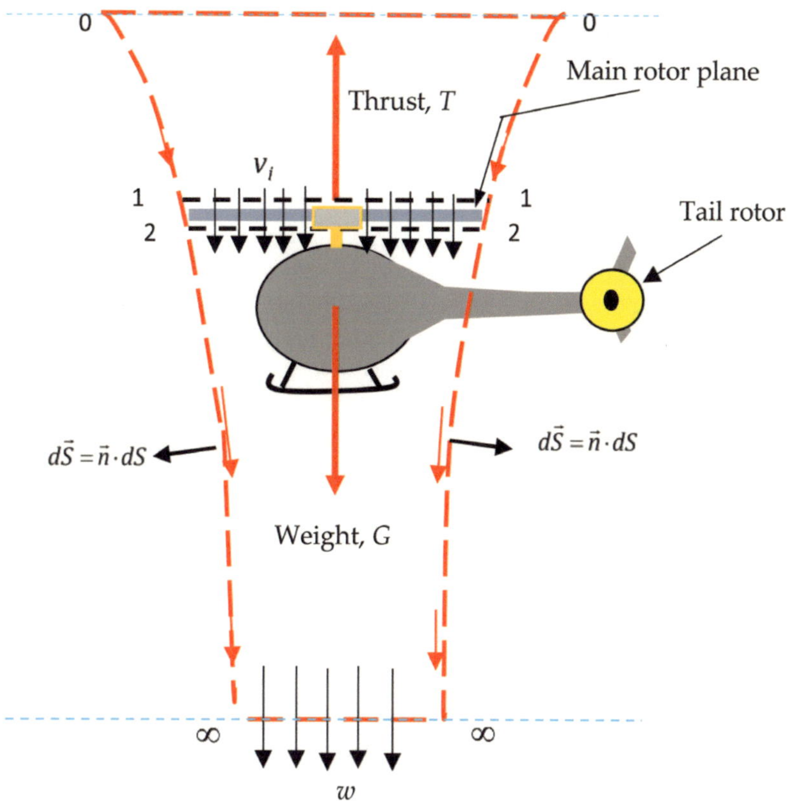

Figure 5. The helicopter in hovering flight.

disk (sections 1–1, and 2–2); the far wake section, denoted by ∞. At the plane of rotor, the velocity through the rotor disk is v_i (named the induced velocity) and in the far wake the air velocity is w. For a control volume surrounding the rotor and its wake, as shown in **Figure 5** and $d\vec{S}=\vec{n}\cdot dS$ the unit normal area vector (the unit normal vector $\vec{n}$ is oriented outward the control volume), according to the Reynolds Transport Theorem, for any extensive parameter B, where $B=b\cdot m$, the following equation is valid

$$\left(\frac{dB}{dt}\right)_{system}=\frac{\partial}{\partial t}\iiint\limits_{\substack{control\\volume}}\rho b dV+\iint\limits_{\substack{control\\surface}}(\rho b)\vec{V}\cdot d\vec{S} \tag{1}$$

where $\vec{V}$ is the local velocity, m is the mass of fluid, and ρ is the fluid density. For a steady flow, the above equation becomes

$$\left(\frac{dB}{dt}\right)_{system}=\iint\limits_{\substack{control\\surface}}(\rho b)\vec{V}\cdot d\vec{S} \tag{2}$$

The conservation of mass (this case corresponds to $B = m$ and $b = 1$)

$$\left(\frac{dm}{dt}\right)_{system} = \iint\limits_{\substack{control\\surface}} (\rho)\vec{V}\cdot d\vec{S} \tag{3}$$

This equation requires the condition that the total amount of mass entering a control volume equals the total amount of mass leaving it. For steady-flow processes, we are not interested in the amount m of mass that flows in or out the control volume, but we are interested in amount of mass flowing per unit time, that is the mass flow rate, $\dot{m}$, well the conservation of fluid mass applied to this finite control volume can be rewritten as

$$-\iint\limits_{surface2} \rho v_i dS + \iint\limits_{surface\infty} \rho w dS = 0 \tag{4}$$

Therefore,

$$\rho v_i A = \rho w A_\infty \tag{5}$$

The conservation of fluid momentum (this case corresponds to $B = m\vec{V}$ and $b = \vec{V}$)

$$\left(\frac{dm\,\vec{V}}{dt}\right)_{system} = \iint\limits_{\substack{control\\surface}} \left(\rho\vec{V}\right)\vec{V}\cdot d\vec{S} \tag{6}$$

The principle of conservation of fluid momentum gives the relationship between the rotor thrust and the time rate of change of fluid momentum out of the control volume. The left part of Eq. (6) represent the sum of all forces that operate upon the control volume, namely the helicopter rotor thrust force, T. In projection on rotational axis, Eq. (6) becomes

$$T = w \iint\limits_{surface\infty} (\rho w) dS = w\dot{m} \tag{7}$$

where $\dot{m}$ is the mass flow rate in the control volume.

The conservation of energy (in this case $B = E = \frac{1}{2}mV^2$ and $b = \frac{1}{2}V^2$)

$$\left(\frac{dE}{dt}\right)_{sistem} = \iint\limits_{\substack{control\\surface}} \left(\rho\frac{1}{2}V^2\right)\vec{V}\cdot d\vec{S} \tag{8}$$

The work done on the helicopter rotor is equal to the gain in energy of the fluid per unit time, and dE/dt represents the power consumed by the rotor, being equal to $T \cdot v_i$, therefore,

$$T \cdot v_i = \iint_{\substack{control \\ surface}} \left(\rho \frac{1}{2} V^2\right) \vec{V} \cdot \vec{dS} = \frac{1}{2} w^2 \dot{m} \tag{9}$$

Taking into account that $T = \dot{m} w$, we have $\dot{m}\, w v_i = \frac{1}{2} w^2 \dot{m}$ or $v_i = \frac{1}{2} w$.

From the equation of continuity $\rho v_i A = \rho w A_\infty$, it follows that $A_\infty = \frac{1}{2} A$ and obviously, $r_\infty = \frac{R}{\sqrt{2}}$ therefore, the ratio of the rotor to the radius of the wake is $R/r_\infty = \sqrt{2}$.

Replacing the velocity w in the vena contracta (section ∞) in the expression of thrust force T, it follows that

$$T = \dot{m} w = \dot{m}(2v_i) = \rho A v_i (2v_i) = 2\rho A v_i^2 \tag{10}$$

The induced velocity at the plane of the rotor disk is v_{hover},

$$v_h = v_i = \sqrt{\frac{T}{A} \frac{1}{2\rho}} \tag{11}$$

This expression shows that induced velocity is dependent explicitly on the disk loading T/A, which is an important parameter in the helicopter design.

The power required to hover is the product between thrust T and induced velocity v_i,

$$P = T \cdot v_i = T \sqrt{\frac{T}{A} \frac{1}{2\rho}} = \frac{T^{\frac{3}{2}}}{\sqrt{2\rho A}} \tag{12}$$

This power, called the ideal power, forms the majority of the power consumed in hover, which is itself a high power-consuming helicopter flight regime.

In assessing rotor performance and compare calculations for different rotors, nondimensional quantities are useful. The induced velocity is normalized using the rotor tip speed, $R\Omega$, where R is the rotor radius and Ω is the angular velocity,

$$\lambda_h = \frac{v_i}{R\Omega} \tag{13}$$

The parameter λ_h is called the induced inflow ratio in hover.

The thrust force is also normalized like the lift for the fixed-wing, that is, the product of a pressure and an area, where the pressure is the dynamic pressure, considered at the rotor blade tips and the area is the total disk area, $A = \pi R^2$, so, the thrust coefficient is defined by

$$C_T = \frac{T}{\frac{1}{2}\rho(R\Omega)^2 \cdot A} \tag{14}$$

The inclusion on the half in the denominator is consistent with the lift coefficient definition for a fixed-wing aircraft. The rotor power, C_P, and rotor torque, C_Q, are defined as

$$C_P = \frac{P}{\frac{1}{2}\rho(R\Omega)^3 \cdot A}; \quad C_Q = \frac{P}{\frac{1}{2}\rho(R\Omega)^2 \cdot R \cdot A} \tag{15}$$

Taking into account that power is related to torque by $P = \Omega \cdot Q$, then numerically $C_P = C_Q$.

Starting from the definition of the induced inflow ratio in hover, λ_h, it follows that

$\lambda_h = \frac{v_i}{R\Omega} = \frac{1}{R\Omega}\sqrt{\frac{T}{2\rho A}} = \sqrt{\frac{T}{4 \cdot \frac{1}{2}\rho A (R\Omega)^2}} = \frac{1}{2}\sqrt{C_T}$, therefore $C_T = 4\lambda_h^2$.

The rotor power coefficient can be represented as

$$C_P = \frac{T \cdot v_i}{\frac{1}{2}\rho(R\Omega)^3 \cdot A} = \frac{T}{\frac{1}{2}\rho(R\Omega)^2 \cdot A}\frac{v_i}{(R\Omega)} = C_T \cdot \lambda_i, \text{ or } C_P = \frac{1}{2}C_T^{\frac{3}{2}} \tag{16}$$

2.2. Vertical climb

Considering the helicopter in climb, one can see that the flow enters the stream tube far upstream of the rotor and then passes through the rotor itself, finally passing away from the rotor forming the wake (**Figure 6**). When the helicopter leaves the hovering condition and moves in a vertical direction, the flow remains symmetrical about the thrust force line, which is normal to the rotor disk. The flow becomes very complex in a medium descent rate condition, but in climb, the mathematical approach is close to that used in the hover conditions.

The air enters the stream tube with velocity V_c and then acquires an additional velocity v_i as it passes through the helicopter rotor disk, and finally, it forms the wake with a velocity $V_c + v_i$. Applying the principles of conservation for mass, momentum, and energy like in the hover we get:

$$\dot{m} = \rho A(V_C + v_i); \quad T = \dot{m}w; \quad w = 2v_i \tag{17}$$

Therefore, $T = \dot{m}w = \rho A(V_C + v_i) \cdot 2v_i$ and dividing by $2\rho A$ it follows that

$$\frac{T}{2\rho A} = (V_C + v_i)v_i = V_C \cdot v_i + v_i^2 \tag{18}$$

The left part of the above equation represents the square of induced velocity in hover, v_h^2, and replacing it, we get

$$v_h^2 = V_C \cdot v_i + v_i^2 \quad \text{or} \quad \left(\frac{v_i}{v_h}\right)^2 + \frac{V_C}{v_h} \cdot \left(\frac{v_i}{v_h}\right) - 1 = 0 \tag{19}$$

The ratio v_i/v_h must always be positive in the climb, so the valid solution is

http://dx.doi.org/10.5772/intechopen.71516

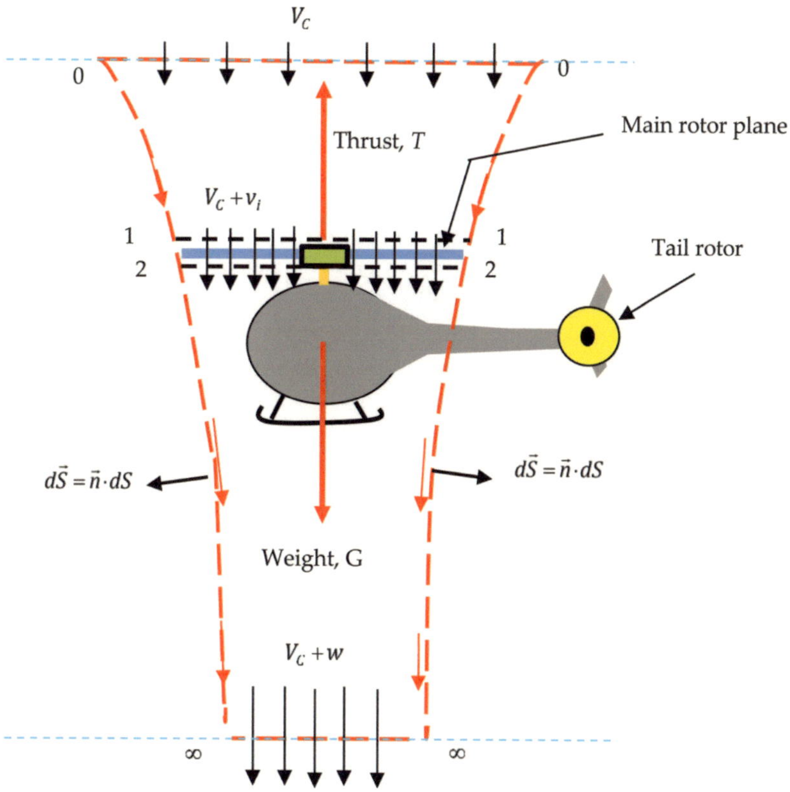

Figure 6. The axial climbing flight.

$$\frac{v_i}{v_h} = -\frac{1}{2}\frac{V_C}{v_h} + \sqrt{\frac{1}{4}\left(\frac{V_C}{v_h}\right)^2 + 1} \tag{20}$$

The power consumed is given by the product of the thrust and the total velocity through the rotor disk, that is

$$P = T(V_c + v_i) = T \cdot V_C + T \cdot v_i = P_{c\,\lim b} + P_i \tag{21}$$

2.3. Vertical descent

In the vertical descent, the air enters the stream tube from below the rotor with velocity V_D and passes through the rotor disk with the velocity $V_D - v_i$, the wake being formed with velocity $V_D - w$, as it is shown in **Figure 7**. The mass flow rate in vertical descent is $\dot{m} = \rho A(V_D + v_i)$, where V_D is negative, and the conservation of momentum gives the thrust force

$$T = \iint_{\substack{control \\ surface}} \left(\rho\vec{V}\right)\vec{V} \cdot d\vec{S} = -\dot{m} \cdot w \tag{22}$$

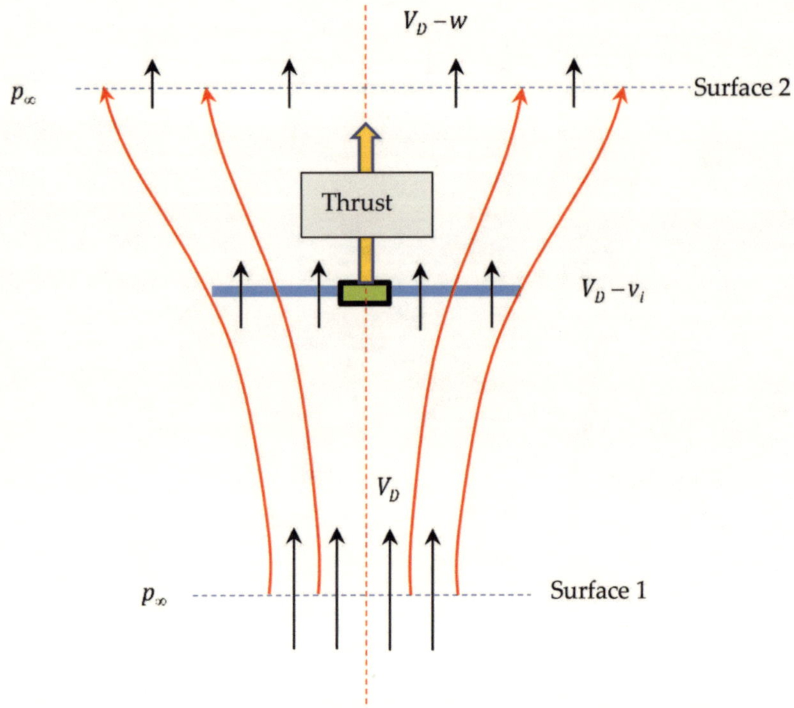

Figure 7. The stream tube in descent.

Even if the sign of thrust is negative, that does not mean that the thrust is negative, because the assumed sign convention consists of positive velocity w, in down direction. According to the conservation energy principle, it follows that

$$T \cdot (V_D - v_i) = -\frac{1}{2}\dot{m}w(2V_D - w) \tag{23}$$

Replacing the expression of thrust T, namely $T = -\dot{m}w$, in the above equation, we have

$$-\dot{m}w(V_D - v_i) = -\frac{1}{2}\dot{m}w(2V_D - w) \tag{24}$$

therefore, $v_i = \frac{w}{2}$.

Similarly, to climb case, having the expression of the mass flow rate $\dot{m} = \rho A(V_D + v_i)$, where velocity V_D is negative and v_i is positive, we can write

$$T = -\dot{m}w = -\rho A(V_D + v_i) \cdot 2v_i = -2\rho A(V_D + v_i)v_i \tag{25}$$

so

$$\frac{T}{2\rho A} = -V_D \cdot v_i - v_i^2 \tag{26}$$

Dividing by $v_h^2 = \frac{T}{2\rho A}$ the above equation becomes

$$\left(\frac{v_i}{v_h}\right)^2 + \frac{V_D}{v_h}\left(\frac{v_i}{v_h}\right) + 1 = 0 \tag{27}$$

with the solutions

$$\frac{v_i}{v_h} = -\frac{1}{2}\frac{V_D}{v_h} \pm \sqrt{\frac{1}{4}\left(\frac{V_D}{v_h}\right)^2 - 1} \tag{28}$$

In order to have real solutions, the following condition must be accomplished

$$\frac{1}{4}\left(\frac{V_D}{v_h}\right)^2 - 1 \geq 0 \tag{29}$$

That means, $|V_D| > 2v_h$. The valid solution is

$$\frac{v_i}{v_h} = -\frac{1}{2}\frac{V_D}{v_h} - \sqrt{\frac{1}{4}\left(\frac{V_D}{v_h}\right)^2 - 1} \tag{30}$$

In the region of flight that corresponds to $-2 \leq V_D/v_h \leq 0$, the control volume cannot be defined and the velocity curve can be defined experimentally. An approximation of the velocity in this region, called vortex ring state, could be [1]

$$\frac{v_i}{v_h} = k + k_1\left(\frac{V_D}{v_h}\right) + k_2\left(\frac{V_D}{v_h}\right)^2 + k_3\left(\frac{V_D}{v_h}\right)^3 + k_4\left(\frac{V_D}{v_h}\right)^4 \tag{31}$$

with $k = 0.974$, $k_1 = -1.125$, $k_2 = -1.372$, $k_3 = -1.718$, and $k_4 = -0.655$.

Figure 8 shows the graphical results from this analysis, made in the Maple soft program.

In the normal working state of the rotor, if the climb velocity increases, the induced velocity decreases and also, in the windmill brake state if the descent velocity increases the induced velocity decreases and asymptotes to zero at high descent rates. In the vortex ring region, the induced velocity is approximated, because momentum theory cannot be applied. The flow in this region is unsteady and turbulent having upward and downward velocities. During normal powered flight, the rotor generates an induced airflow going downward and there is a recirculation of air at the blade tips, having the form of vortices, which exist because higher pressure air from below the rotor blade escapes into the lower pressure area above the blade. The rate of descent that is required to get into the vortex ring state varies with the speed of the induced airflow. Although vortices are always present around the edge of the rotor disk, under certain airflow conditions, they will intensify and, coupled with a stall spreading outward from the blade root, result in a sudden loss of rotor thrust. Vortex ring can only occur when the following conditions are present: power on, giving an induced flow down through rotor disk; a rate of descent, producing an external airflow directly opposing the induced flow; low forward speed.

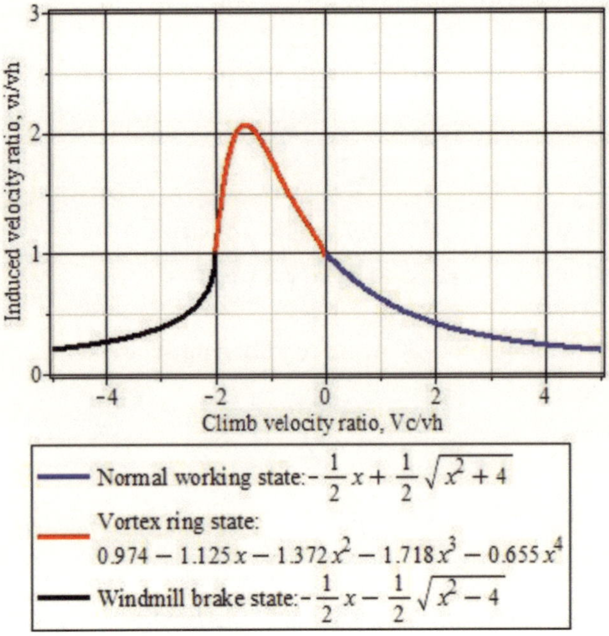

Figure 8. Induced velocity variation.

2.4. Power required in axial climbing and descending flight

In a climb or descent, the power ratio is

$$\frac{P}{P_h}=\frac{V_{C,D}+v_i}{v_h}=\frac{V_{C,D}}{v_h}+\frac{v_i}{v_h} \tag{32}$$

Using Eqs. (20) and (30), and substituting in the above equation, it follows that

- For a climb: $\frac{P}{P_h}=\frac{1}{2}\frac{V_C}{v_h}+\sqrt{\frac{1}{4}\left(\frac{V_C}{v_h}\right)^2+1}$, which is valid for $\frac{V_C}{v_h}\geq 0$;
- For a descent: $\frac{P}{P_h}=\frac{1}{2}\frac{V_D}{v_h}-\sqrt{\frac{1}{4}\left(\frac{V_D}{v_h}\right)^2-1}$, which is valid for $\frac{V_C}{v_h}\leq -2$;

For the vortex ring state, we can use the approximation (31) for the induced velocity ration, therefore in this case, the power ratio is

$$\frac{P}{P_h}=\frac{V_D}{v_h}+\frac{v_i}{v_h}=\frac{V_D}{v_h}+k+k_1\left(\frac{V_D}{v_h}\right)+k_2\left(\frac{V_D}{v_h}\right)^2+k_3\left(\frac{V_D}{v_h}\right)^3+k_4\left(\frac{V_D}{v_h}\right)^4 \tag{33}$$

Using the same Maple soft program like for induced velocity, we obtain the following picture for the power ratio, P/P_h.

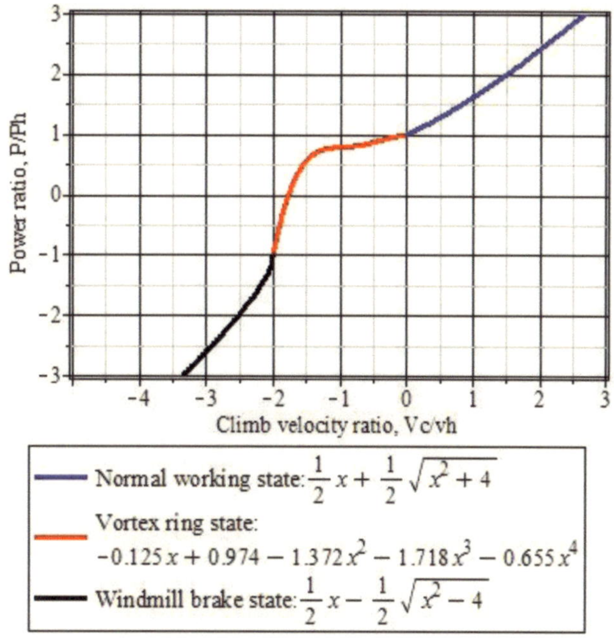

Figure 9. Power required as a function of climb and descent velocity.

According to the power to power in hover ratio values, shown in **Figure 9**, the power required to climb is always greater than the power required to hover, namely this ratio is greater than unity. In descent flight, the rotor extracts power from the air and uses less power than to hover.

2.5. Induced velocity in forward flight

In forward flight, the rotor must be tilted (**Figure 10**) in order to have a propulsive force to propel the helicopter forward, with a velocity V_∞. This velocity has two components: one component normal to the rotor disk plane, $V_\infty \sin\alpha$, and another component, parallel to the rotor disk plane, $V_\infty \cos\alpha$.

The rotor thrust, T, is given by $T = 2\rho A v_i \sqrt{(V_\infty \cos\alpha)^2 + (V_\infty \sin\alpha + v_i)^2}$ and the induced velocity in forward flight can be written as

$$v_i = \frac{\frac{T}{2\rho A}}{\sqrt{(V_\infty \cos\alpha)^2 + (V_\infty \sin\alpha + v_i)^2}} = \frac{v_h^2}{\sqrt{(V_\infty \cos\alpha)^2 + (V_\infty \sin\alpha + v_i)^2}} \tag{34}$$

In order to get an analytical solution for the induced velocity, it is necessary to define two coefficients: *the advance ratio*, μ, and *the inflow ratio*, λ, as it follows

$$\mu = \frac{V_\infty \cos\alpha}{R\Omega} \tag{35}$$

$$\lambda = \frac{V_\infty \sin\alpha + v_i}{R\Omega} = \frac{V_\infty \sin\alpha}{R\Omega} + \frac{v_i}{R\Omega} = \mu \tan\alpha + \lambda_i \tag{36}$$

Dividing Eq. (34) to $R\Omega$, we get

$$\frac{v_i}{R\Omega} = \lambda_i = \frac{\frac{v_h^2}{(R\Omega)^2}}{\sqrt{\left(\frac{V_\infty \cos\alpha}{R\Omega}\right)^2 + \left(\frac{V_\infty \sin\alpha + v_i}{R\Omega}\right)^2}} = \frac{\lambda_h^2}{\sqrt{\mu^2 + \lambda^2}} \tag{37}$$

This expression leads to the following equation for the inflow ratio, λ,

$$\lambda = \mu \tan\alpha + \frac{\lambda_h^2}{\sqrt{\mu^2 + \lambda^2}}, \text{ or } \frac{\lambda}{\lambda_h} = \frac{\mu}{\lambda_h} \tan\alpha + \frac{1}{\sqrt{\left(\frac{\mu}{\lambda_h}\right)^2 + \left(\frac{\lambda}{\lambda_h}\right)^2}} \tag{38}$$

The above equation can be very easy to be solved in Maple soft. In **Figure 11**, three curves are shown, in coordinates $\frac{\mu}{\lambda_h}$ and $\frac{\lambda}{\lambda_h}$, representing three values of angle α, namely $\alpha = 0$ deg, $\alpha = 4$ deg, and $\alpha = 6$ deg. The command plot used in Maple was "*implicitplot*" for the equation $y = x \cdot \tan\alpha + \frac{1}{\sqrt{x^2 + y^2}}$.

The inflow ratio, λ/λ_h increases with the increase of the rotor disk angles of attack.

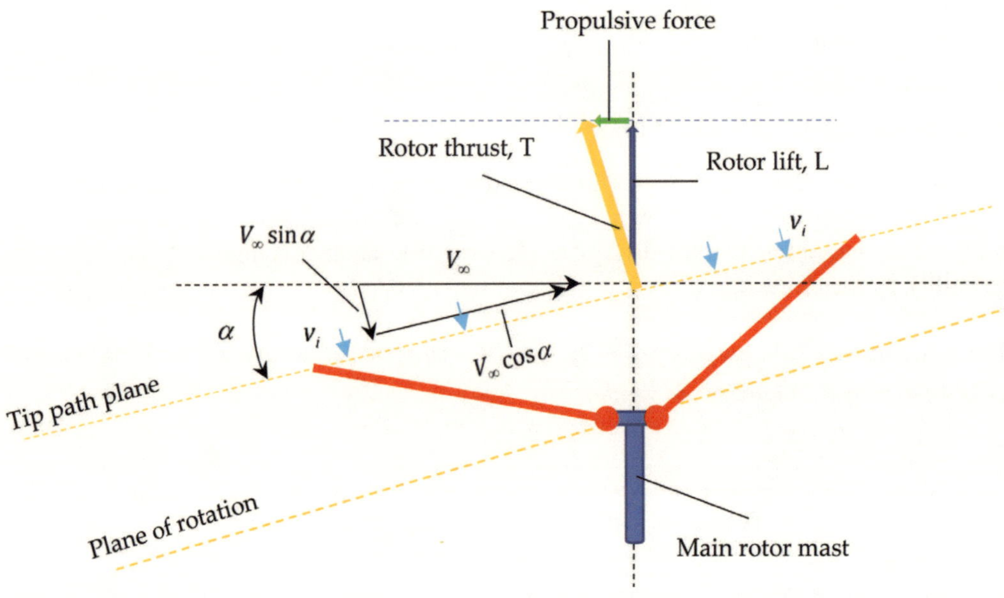

Figure 10. Rotor in forward flight.

http://dx.doi.org/10.5772/intechopen.71516

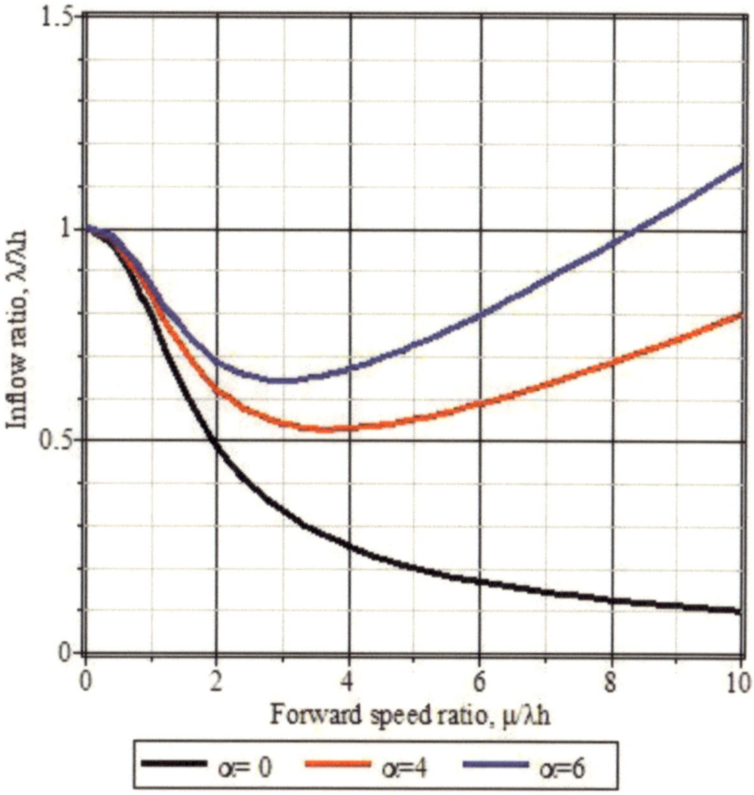

Figure 11. Inflow ratio λ/λ_h as a function of forward speed ratio.

3. Helicopter systems

3.1. Main rotor systems

The primary way to distinguish between different main rotor systems is represented by the movement of the blade relative to the main rotor hub. The main categories are fully articulated, semi rigid, and rigid. In hovering flight, the blades flap up and lag back with respect to the hub and reach equilibrium position under the action of aerodynamic and centrifugal forces. In forward flight, the asymmetry of the dynamic pressure over the disk produces aerodynamic forces that are the functions of the blade azimuth position. The hinges allow each blade to independently flap and lead or lag with respect to the hub plane. The lead-lag hinge allows in-plane motion of the blade due to the Coriolis and radius of gyration changing in flapping movement. Transition from hover to forward flight introduces additional aerodynamic forces and effects that are not found when the helicopter is in stationary hover. Due to the difference in relative airspeed between the advancing and retreating blades, the lift is constantly changing through each revolution of the rotor.

Figure 12 shows the flapping, lead-lag, and feathering motion of a rotor blade.

In a fully articulated rotor, each main rotor blade is free to move up and down (flapping), to move forth and back (dragging), and to twist about the spanwise axis (feathering). Semi rigid rotor has, normally, two blades attached rigidly to the main rotor hub and is free to tilt and rock independently of the main rotor mast, one blade flaps up and other flaps down.

The rigid rotor system cannot flap or drag, but it can be feathered. The natural frequency of the rigid rotor is high, so the stability is difficult to be achieved.

3.2. Anti-torque system

The single rotor helicopters require a separate rotor to overcome the effect of torque reaction, namely the tendency for the helicopter to turn in the opposite direction to that of the main rotor. The anti-torque pedals are operated by the pilot's feet and vary the force produced by the tail rotor to oppose torque reaction.

3.3. Swash plate assembly

It has the purpose to transmit cyclic and collective control movements to the main rotor blades and consists of a stationary plate and a rotating plate. The stationary plate is attached to the

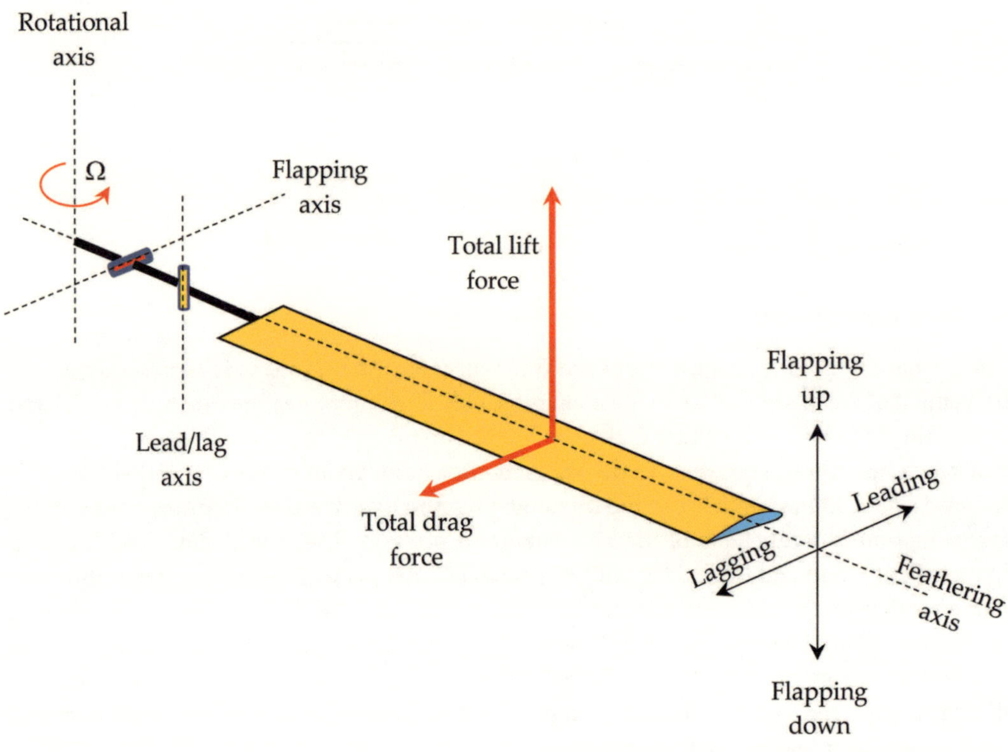

Figure 12. Blade movement axis.

main rotor mast and the rotating plate is attached to the stationary plate by a bearing surface and rotates at the same speed as the main rotor blades.

3.4. Trim

The neutral position of the cyclic stick changes as the helicopter moves off from to hover in forward flight. Trim control can adjust the mechanical feel in flight by changing the neutral position of the stick.

3.5. Collective and cyclic pitch control

Collective pitch lever controls the lift produced by the rotor, while the cyclic pitch controls the pitch angle of the rotor blades in their cyclic rotation. This tilts the main rotor tip-path plane to allow forward, backward, or lateral movement of the helicopter.

4. Power

4.1. Power required

The power required for flight is the second work that must be transmitted to the shaft of the rotor.

In general, for a helicopter in forward flight, the total power required at the rotor, P, can be expressed by the equation

$$P = P_i + P_o + P_P + P_y \tag{39}$$

where P_i is the induced power, P_0 is the profile power required to overcome viscous losses at the rotor, P_P is the parasitic power required to overcome the drag of the helicopter, and P_y is the climb (or descend) power required to increase (decrease) the gravitational potential of the helicopter [1].

Inductive power is consumed to produce lift equal to the weight of the helicopter. From the simple 1-D momentum theory the induced power of the rotor, P_i, can be approximated as

$$P_i = k \cdot T \cdot v_i \tag{40}$$

where k is the familiar empirical correction to account for a multitude of aerodynamic phenomena, mainly those resulting from tip losses and nonuniform inflow, and v_i is induced velocity [1].

The profile power required to overcome the profile drag of the blades of the blades of the rotor is

$$P_0 = Q \cdot \Omega \tag{41}$$

where Q is the rotor torque, and Ω is the rotational frequency of the rotor.

The parasite power, P_P, is a power loss as a result of viscous shear effects and flow separation (pressure drag) on the fuselage, rotor hub, and so on. Because helicopter fuselages are much less aerodynamic than their fixed-wing counterparts (for the same weights), this source of drag can be very significant [1]. The parasite power can be written as

$$P_P = D \cdot V \tag{42}$$

The climb (or descend) power can be written as

$$P_y = T(v_i \pm V_y) \tag{43}$$

where V_y is the climb (or descend) velocity. In hover regime $V_y=0$.

In addition, when calculating the power required of the helicopter, the required power of the tail rotor must also be calculated. The power required by the tail rotor typically varies between 3 and 5% of the main rotor power in normal flight, and up to 20% of the main rotor power at the extremes of the flight envelope [1]. It is calculated in a similar way to the main rotor power, with the thrust required being set equal to the value necessary to balance the main rotor torque reaction on the fuselage. The use of vertical tail surfaces to produce a side force in forward flight can help to reduce the power fraction required for the tail rotor, albeit at the expense of some increase in parasitic and induced drag.

Figure 13 shows the net power required for a given helicopter in straight-and-level flight.

4.2. Power available

The power needed to rotate the main rotor transmits to the main rotor from the engine through the transmission (**Figure 13**). But the main rotor cannot get all the power, which is developed from the engine, as part of it is spent for other purposes and does not go to the main rotor.

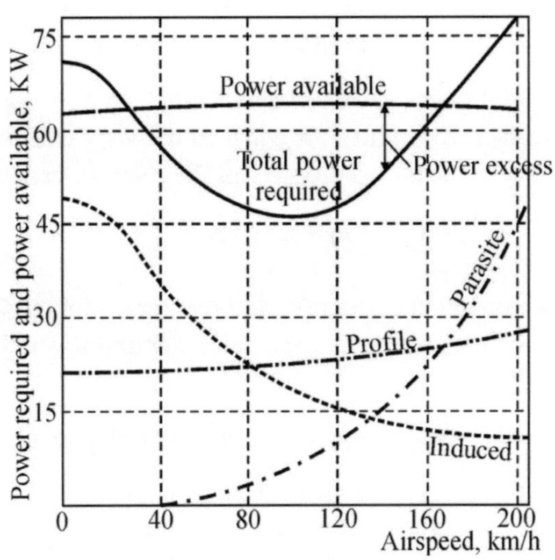

Figure 13. Power required and power available in straight-andlevel flight.

For rotating of the tail rotor about 8% is lost from the consumed power of the engine, for fan rotation about 5%, for friction about 7% in transmission, for auxiliary drive units about 1%, and for blowing parts of the helicopter about 2%.

This part of the power of the motor that is transmitted to the main rotor is called available power. It is defined as the difference between effective power and total loss.

Excess power—this is the difference between the available and the power required. The greater the excess power is, the greater the speed range is and the better the helicopter's maneuvering characteristics are (**Figure 13**).

5. Ground effect

When the helicopter flies near the earth's surface, the efficiency of the rotor system increases because of the interference of the airflow with the ground [2–4]. The rotor downwash is unable to escape as readily as it can when flying higher and creates a ground effect. When the rotor downwash reaches the surface, the induced flow downwash stops its vertical velocity, which reduces the induced flow at the rotor disk (**Figure 14**).

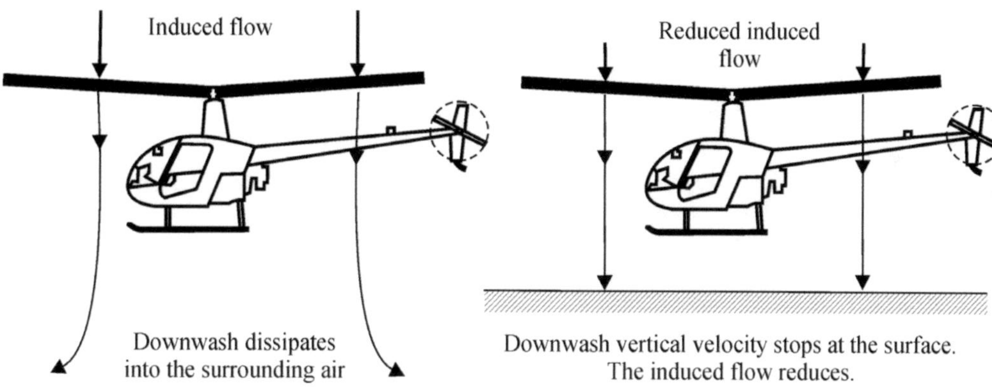

Figure 14. Influence of ground effect on the induced flow.

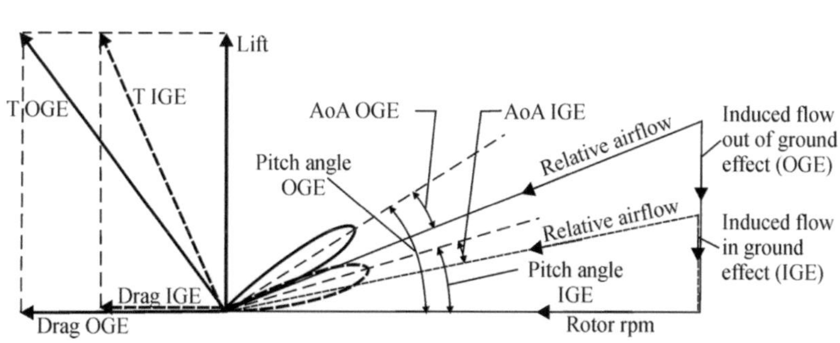

Figure 15. Influence of ground effect on the rotor drag.

Figure 15 shows the effects of this on the power required to hover. If the hover height in ground effect must be maintained, the aircraft can only be kept at this height by reducing the angle of attack (AoA) so that the total reaction produces a rotor lift exactly equal and opposite to weight. It shows that the angle of attack is slightly less, the amount of total rotor thrust is the same as the gross weight, the blade angle is smaller, the power required to overcome the reduced rotor drag (or torque) is less and the collective control lever is lower than when hovering out of ground effect. Therefore, there is better lift/drag ratio.

These conclusions are also true to flight in ground effect other than the hover, but the effect is smaller.

6. Autorotation

Autorotation is an emergency mode. It can arise if the engine stops in flight (usually without the pilot's desire), when the rotor is not driven by the engine and begins to rotate by aerodynamic forces resulting from rate of oncoming airflow through the rotor [1, 3].

6.1. Vertical autorotation

In the case of vertical autorotative descent (without forward speed) without wind, the forces that cause a rotation of the blades are similar for all blades, regardless of their azimuth position [2].

During vertical autorotation, the rotor disk is divided into three regions (as illustrated in **Figure 16a**): driven region, driving region, and stall region. **Figure 17** shows the blade sections that illustrate force vectors. Force vectors are different in each region, as the relative air velocity is lower near the root of the blade and increases continually toward its tip. The combination of the inflow up through the rotor with the relative air velocity creates different aerodynamic forces in each section along the blade [2].

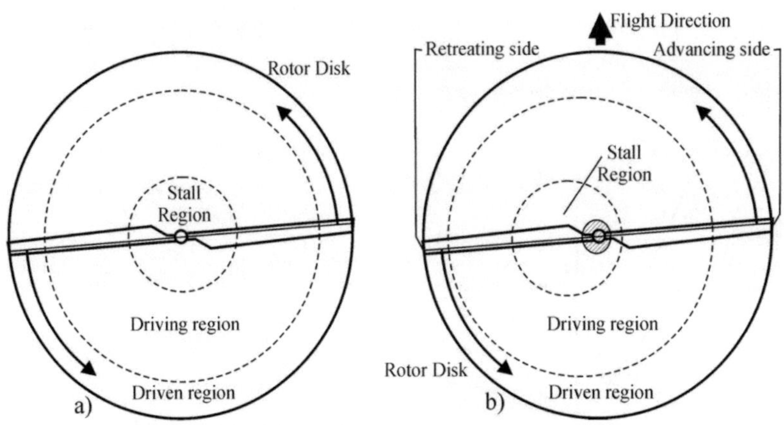

Figure 16. Autorotation regions in (a) vertical descend and (b) forward autorotation descend.

http://dx.doi.org/10.5772/intechopen.71516

In the driven region, illustrated in **Figure 17**, the section aerodynamic force T acts behind the axis of rotation. This force has two projections: the drag force D and lift force L. In this region, the lift is offset by drag, and the result is a deceleration of the blade rotation. There are two sections of equilibrium on the blade—the first is between the driven area and the driving region, and the second is between the driving region and the stall region. At the equilibrium sections, the aerodynamic force T coincides with the axis of rotation. There are lift and drag forces, but neither acceleration nor deceleration is induced [2].

In the driving region, the blade produces the forces needed to rotate the blades during the autorotation. The aerodynamic force in the driving region is inclined slightly forward with respect to the axis of rotation. This inclination provides thrust that leads to an acceleration of the blade rotation. By controlling the length of the driving region, the pilot can adjust the autorotative rpm [2].

In the stall region, the rotor blade operates above its stall angle (maximum angle of attack), causing drag, which tends to slow rotation of the blade.

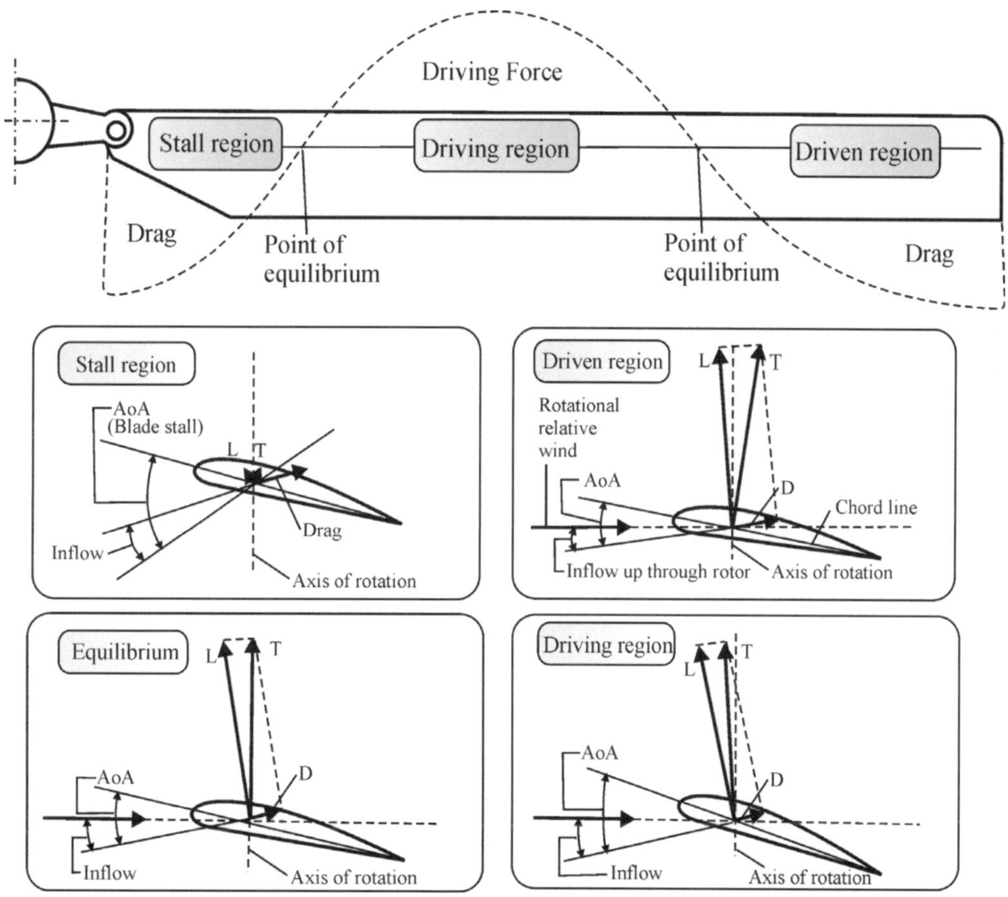

Figure 17. Force vectors in vertical autorotation.

6.2. Autorotation in forward descend

Autorotative force in forward flight is produced in exactly the same scheme as when the helicopter is descending vertically in still air. However, because of the forward flight velocity there is a loss of axial symmetry in the induced velocity and angles of attack over the rotor disk. This tends to move the distribution of parts of the rotor disk that consume power and absorb power, as shown in **Figure 16b**. A small section near the root experiences a reversed flow; therefore, the size of the driven region on the retreating side is reduced [1].

7. Helicopter stability and control

7.1. Helicopter stability

Helicopter stability means its ability in the conditions of external disturbances to keep the specified flight regime without pilot management [3, 5].

Let us consider the longitudinal motion of a helicopter on the hovering regime (**Figure 18**). The weight of the helicopter *W*, attached to the helicopter's CG, is balanced by the thrust force of the main rotor *T*, applied at a point removed on the vertical axis by the distance *z* (see **Figure 18a**).

Recall that a helicopter, like any aircraft, is considered statically stable, if it after a deviation from the steady flight regime tends to return to its original position. Suppose, for example, that as a result of the action of a wind gust *U* the thrust *T* is deflected backward (see **Figure 18b**). The tilt of the thrust will result in the appearance of a horizontal component *D* acting backwards and a longitudinal pitching moment $M = Dz$. Under the action of the horizontal component, the helicopter will start to move back with a speed V_x, and under the action of the moment *M* it will start to rotate relative to the roll axis, increasing the pitch angle with the angular velocity *q* (see **Figure 18c**).

Both effects: both the translational velocity and the rotation of the fuselage, and hence the axis of the rotor, will cause the resultant forces *T* on the rotor to tilt to the same side, opposite to the original inclination. This will cause the appearance of a horizontal component and a longitudinal moment, already oppositely directed, due to which the helicopter will tend to return to the initial pitch angle and to zero forward speed. This means that the helicopter is statically stable in pitch angle and hover speed. Its static stability is due to the properties mentioned above: speed stability and damping.

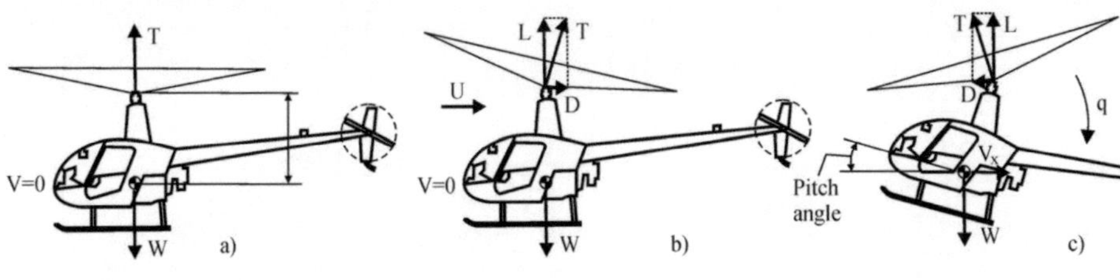

Figure 18. Longitudinal motion of the helicopter in hover.

Consider, however, the further movement of the helicopter. The inclination of the resultant in the direction of parrying disturbance is too great because of the presence of velocity stability. It leads to the fact that the helicopter in its movement to the initial position skips the equilibrium position and deviates in the opposite direction, but already by a large magnitude. The motion of the helicopter takes the character of oscillation with increasing amplitude. The aircraft, which in the free disturbed motion ultimately leave the initial equilibrium state, is called dynamically unstable. Thus, a helicopter on a hovering regime is dynamically unstable.

The given case relates to the helicopter's movement on the pitch angle on the hover. The roll motion on the hover has a similar character. The difference here is manifested only in the period and the degree of growth of oscillation, which depend on the moments of inertia of the helicopter, different in pitch and roll.

The helicopter is neutral in the yaw angle and the altitude on the hover. This means that the helicopter does not tend to keep a given course angle or a given flight altitude. At the corresponding disturbances these parameters will change. But their change will continue only as long as the perturbation is working. At the end of the disturbance, the course angle and altitude will not change.

It can be said that the helicopter is stable with respect to the yaw rate and the vertical speed. This stability is explained by the fact that the main rotor at an increase of the airspeed in a direction opposite to the thrust reduces its thrust, and conversely, when this speed decreases—increases the thrust, thus creating a damping force in the direction of the axis of rotation. Therefore, the tail rotor creates a large damping yaw moment on the helicopter, and the main rotor—a damping force for vertical helicopter movements.

In forward flight, the efficiency of helicopter control and the derivatives of the damping moments and moments of stability with respect to the main rotor speed vary insignificantly. However, the moment derivative with respect to the angle of attack, which for the main rotor corresponds to the instability, begins to play an important role. This instability can be compensated if the fuselage of the helicopter has a stabilizer, which improves the desired degree of stability in the angle of attack. But it is difficult to provide satisfactory longitudinal stability even with well-designed stabilizer. That's why the modern helicopters are equipped with electronic stabilization.

In the forward flight, the roll movement is strongly connected with the yaw movement, just as it does on the airplane. These two movement types are therefore referred to as one, "lateral" movement of the helicopter. The own lateral motion of a single-rotor helicopter during a forward flight, as a rule, is periodically stable. In the low-speed modes, while the relationship between the roll and yaw movements is still small, and the roll motion, like the hovering, is unstable, the lateral motion of a single-rotor helicopter is unstable.

Static stability of helicopters with two main rotors differs slightly from the stability of the helicopter with one main rotor. The tandem main rotor helicopter has a significantly greater longitudinal static stability, and the coaxial main rotor helicopter has a greater lateral stability. This is explained by the change of main rotors thrust at a disruption of the equilibrium.

So, the helicopter, essentially, cannot maintain a steady flight regime. The pilot, piloting the helicopter, continuously has to act on the helicopter's controls and create control moments, under which the helicopter to maintain the specified flight regime.

7.2. Helicopter control

Control characteristics refer to a helicopter's ability to respond to control inputs and so move from one flight condition to another [6]. There are four basic controls used during flight. They are the collective pitch control, the throttle, the cyclic pitch control, and the antitorque pedals (**Figure 19**).

7.3. Collective pitch control

The collective pitch control changes the pitch angle of all main rotor blades. The collective is controlled by the left hand (**Figure 19**). As the pitch of the blades is increased, lift is created

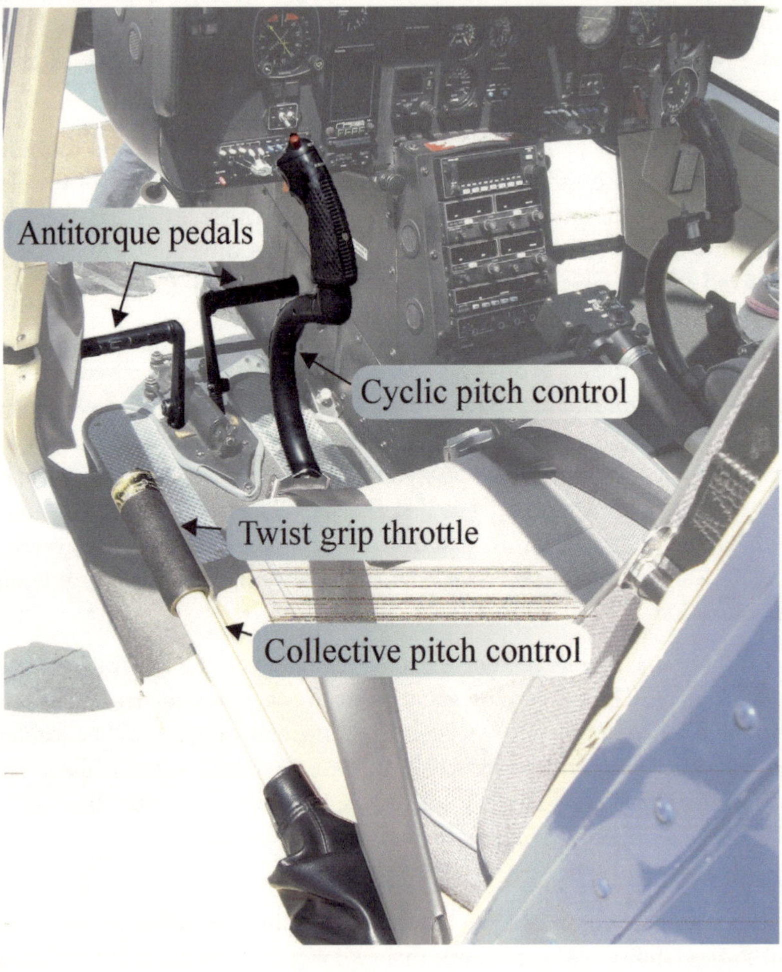

Figure 19. Basic helicopter controls.

causing the helicopter to rise from the ground, hover or climb, as long as sufficient power is available.

The variation of the pitch angle of the blades changes the angle of attack on each blade. The change in the angle of attack causes a change in the drag, which reflects the speed or rpm of the main rotor. When the pitch angle increases, the angle of attack increases too, therefore the drag increases, and the rotor rpm decreases. When the pitch angle decreases, the angle of attack and the drag decrease too, but the rotor rpm increases. To maintain a constant rotor rpm, which is specific to helicopters, a proportional alteration in power is required to compensate for the drag change. This is achieved with a throttle control or a correlator and/or governor, so that the engine power can be regulated automatically [2].

7.4. Throttle control

The purpose of the throttle is to regulate engine rpm if the system with a correlator or governor does not maintain the necessary rpm when the collective is raised or lowered, or if those devices are not installed, the throttle has to be moved manually with the twist grip to maintain desired rpm. Twisting the throttle outboard increases rpm; twisting it inboard decreases rpm [2].

The correlator is a device that connects the collective lever and the engine throttle. When the collective lever raises, the power automatically increases and when lowers, the power decreases. The correlator maintains rpm close to the desired value, but still requires an additional fine tuning of the throttle. The governor is a sensing device that recognizes the rotor and engine rpm and makes the necessary settings to keep rotor rpm constant. Under normal operation, once the rotor rpm is set, the governor keeps the rpm constant, and there is no need to make any throttle settings. The governor is typical device used in turbine helicopters and is also used in some helicopters with piston engines [2].

7.5. Cyclic pitch control

The rotor control is performed by the cyclic pitch control, which tilts the main rotor disk by changing the pitch angle of the rotor blades. The tilting rotor disk produces a cyclic variation of the blade pitch angle. When the main rotor disk is tilted, the horizontal component of thrust moves the helicopter in the tilt direction.

Figure 20 shows the conventional main rotor collective and cyclic controls. The controls use a swash plate. The collective control applies the same pitch angle to all blades and is the main tool for direct lift or thrust rotor control. Cyclic is more complicated and can be fully appreciated only when the rotor is rotating. The cyclic operates through a swash plate (**Figure 20**), which has non-rotating and rotating plates, the latter attached to the blades with pitch link rods, and the former to the control actuators [7].

7.6. Antitorque pedals

Two anti-torque pedals are provided to counteract the torque effect of the main rotor. This is done by increasing or decreasing the thrust of the tail rotor (**Figures 19** and **21**). The torque varies with changes in main rotor power; therefore, the tail rotor thrust is necessary to change

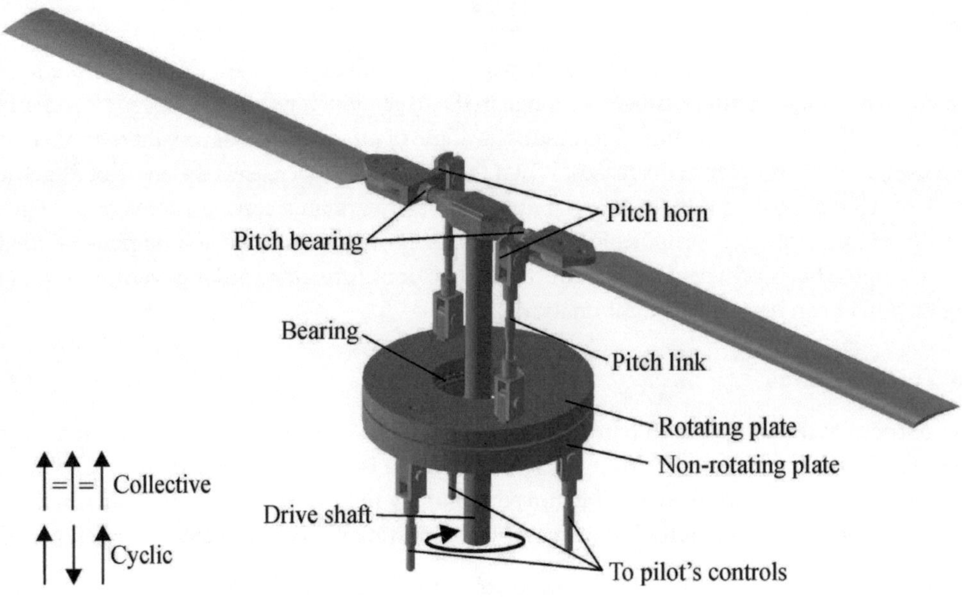

Figure 20. Rotor control through a swash plate.

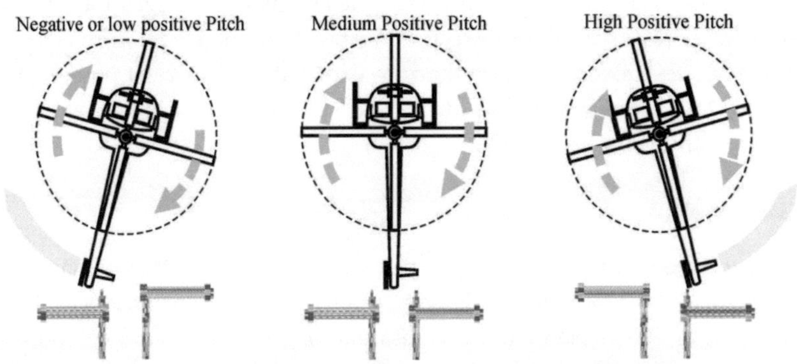

Figure 21. Tail rotor pitch angle and thrust in relation to pedal positions during cruising flight.

too. The pedals are connected to the pitch change device on the tail rotor gearbox and enable the pitch angle of the tail rotor blades to increase or decrease [2].

8. Flight performance

The pilot's ability to determine, in advance, the helicopter's flying characteristics is of utmost importance. It is very important to determine what maximum weight the helicopter can carry before take-off, if the helicopter can safely hover at a given altitude and temperature, what distance is needed to climb above the obstacles, and what is the maximum climb rate [2].

http://dx.doi.org/10.5772/intechopen.71516

8.1. Factors affecting performance

There are many factors that influence a helicopter's performance in flight. The most important ones are: altitude, including pressure altitude and density altitude, helicopter gross weight, and the wind.

8.2. Altitude

One of the most important factors in helicopter performance is the air density, which decreases with a gain in altitude. The effect of altitude is shown in **Figure 22a**. Increasing density altitude increases the power required in hover and lower airspeeds. At higher airspeeds, the results of lower air density result in a lower power requirements because of the reduction of parasitic drag. A higher density altitude also affects the engine power available. The power available at a higher density altitude is less than that at a lower one. As a result there is a decrease in the excess power at any airspeed [1].

8.3. Weight

Increases in aircraft gross weight go hand in hand with requirements for higher angles of attack and more power. As shown in **Figure 22b**, by increasing the weight, the excess power becomes less, but it is particularly affected at lower airspeeds because of induced drag [1].

High gross weight also affects of the maximum height at which the helicopter can operate in ground effect for a given power available. Under these conditions, the heavier the helicopter is, the lower the maximum hover altitude is [3].

8.4. Wind

Wind direction and velocity also affect hovering, takeoff, and climb performance. Translational lift occurs any time when there is relative airflow over the rotor disk. This explains whether the relative airflow is caused by helicopter movement or by the wind. With the increase in the wind speed, the translational lift increases, therefore less power is required in hovering [2].

Besides the magnitude of wind velocity, its direction is essential. Headwind is the most desirable because it gives the greatest increase in performance. Strong crosswind and tailwind require the

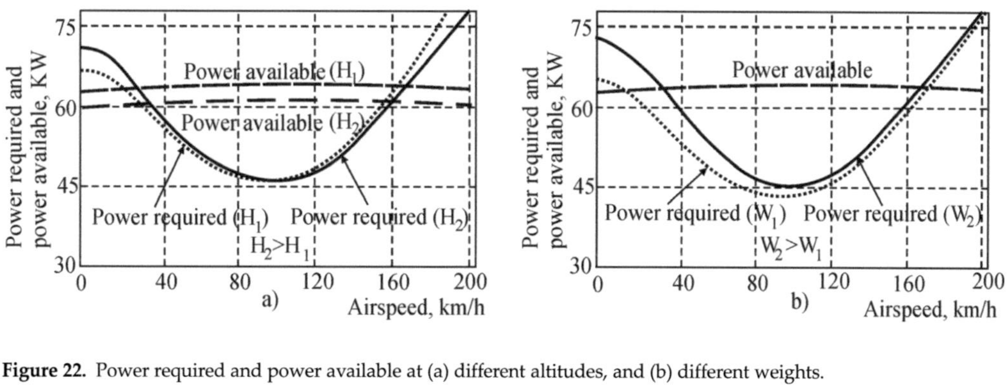

Figure 22. Power required and power available at (a) different altitudes, and (b) different weights.

more tail rotor thrust to maintain the directional control. The increased tail rotor thrust takes away a power from the engine, and therefore will have less power available to the main rotor, which produces the required lift. Some helicopters have a critical wind azimuth limits and the manufacturer presents maximum safe relative wind chart. If the helicopter operates above these limits, it can cause a loss of tail rotor control [2].

8.5. Performance charts

When developing performance charts, aircraft manufacturers make some assumptions about the operating helicopter conditions and the pilot's ability. It is supposed that the helicopter is in good operating condition and the engine is able to develop its rated power. It is assumed that the pilot performs normal operating procedures and he has average flying abilities [2].

With these assumptions, the manufacturer develops performance data for the helicopter taking into account the flight tests. But the helicopter is not tested under all conditions shown on the performance chart. Instead, an evaluation of the specific data is performed and the remaining data are obtained in mathematical way [2].

Generally, the charts present graphics related to hover power: in ground effect (IGE) hover ceiling vs. gross weight, and out of ground effect (OGE) hover ceiling vs. gross weight. The exact names of these charts may vary by different helicopter manuals. These are not the only charts, but these charts are perhaps the most important charts in each manual—they help to understand the amount of power which the helicopter have to have under specific operating conditions (altitude, gross weight, and temperature).

9. Conclusion

In this chapter, an analysis for defining the helicopter's performance was performed. It has been shown that the performance characteristics can be derived by using simple models as the momentum and blade elements theories. The impact of weight and altitude on the required power and the available power has been presented. The effect of near the ground operation on the helicopter's performance is discussed. Also, the case when the engine stops in flight and the main rotor performs autorotation is presented. Some elementary analysis of the stability characteristics has been done. The impact of different helicopter parts on the stability has been considered. Finally, it has been shown how the helicopter can be controlled.

Author details

Constantin Rotaru[1]* and Michael Todorov[2]

*Address all correspondence to: rotaru.constantin@afahc.ro

1 "Henri Coanda" Air Force Academy, Brasov, Romania

2 Technical University of Sofia, Bulgaria

http://dx.doi.org/10.5772/intechopen.71516

References

[1] Leishman G. Principles of Helicopter Aerodynamics. 1st ed. Cambridge: Cambridge University Press; 2000. p. 495

[2] Johnson W. Helicopter Theory. New York: Dover Publications; 1994. p. 1085

[3] Wagtendonk W. Principles of Helicopter Flight. 2nd ed. Newcastle: Aviation Supplies & Academics Inc.; 2015. p. 304

[4] Johnson W. Rotorcraft Aeromechanics. 1st ed. New York: Cambridge University Press; 2013. p. 927

[5] Prouty R. Helicopter Performance, Stability, and Control. 3rd ed. Florida: Kreiger Publishing Company; 1995. p. 731

[6] Seddon J, Newman S. Basic Helicopter Aerodynamics. 3rd ed. Chichester: John Wiley & Sons; 2011. p. 255

[7] Padfield G. Helicopter Flight Dynamics. 2nd ed. Oxford: Blackwell Publishing; 2007. p. 641

Chapter 3

Flight Dynamic Modelling and Simulation of Large Flexible Aircraft

Gaétan Dussart, Vilius Portapas,
Alessandro Pontillo and Mudassir Lone

Additional information is available at the end of the chapter

http://dx.doi.org/10.5772/intechopen.71050

Abstract

The drive for aircraft efficiency and minimum environmental impact is requiring the aerospace industry to generate technologically innovative and highly integrated aircraft concepts. This has changed the approach towards conceptual design and highlighted the need for modular low fidelity aircraft simulation models that not only capture conventional flight dynamics but also provide insight into aeroservoelasticity and flight loads. The key aspects that drive the need for modularity are discussed alongside integration aspects related to coupling aerodynamic models, flight dynamic equations of motion and structural dynamic models. The details of developing such a simulation framework are presented and the utility of such a tool is illustrated through two test cases. The first case focuses on aircraft response to a gust that has a spanwise varying profile. The second investigates aircraft dynamics during control surface failure scenarios. The Cranfield Accelerated Aeroplane Loads Model (CA^2LM) forms the basis of the presented discussion.

Keywords: modelling, simulation, flight dynamics, flexible aircraft, aeroelastic coupling

1. Introduction

Today's concerns regarding growth in the demand for air transport and the environmental impact of aviation has resulted in active efforts by airframe manufacturers to design more efficient aircraft. They have adopted a strategy that sees an incremental introduction of novel technologies, where at each stage the components that constitute the aircraft become more integrated with each other. This effectively provides the opportunity to build the multi-disciplinary design tools and experience needed to develop radical configurations. As a result, the technical disciplines in aircraft design which have traditionally been relatively independent, such as aeroelasticity and flight dynamics, must now integrate. This chapter aims to present the methods used for developing modelling and simulation tools that are needed to facilitate such an integrated approach, especially focusing on large flexible aircraft.

The traditional approach to modelling and simulation of aircraft flight dynamics has framed the problem in the form of the equations of motion (EoM) that couple nonlinear inertial components with quasi-linear aerodynamic models [1, 2]. This has been found to be satisfactory when modelling the flight dynamics of rigid aircraft, but the assumptions of linearity in the method used to formulate the aerodynamic model remains the primary limitation of this approach. Typically, this limitation is the cause of significant uncertainty early in the aircraft design process where engineers can only resort to either empirical methods or panel based methods. For conventional tube and wing configurations, the civil aviation industry has developed and modified these methods based on extensive testing and operational data. On the other hand, the radical configurations seen in the military domain rely on significant effort put towards the identification of aerodynamic characteristics and validation of models during the expensive flight test phase. The latter may often span the entire service life of the aircraft [3, 4].

Accurate modelling and simulation of novel concepts aimed to address today's societal concerns is needed to enable the multidisciplinary approach necessary for design. However, it cannot resort to the knowledge gained either from significant operational data or extensive flight test data. As a result it can only rely on a physics based approach and moreover, this approach needs to be modular if it is to assist in the necessary multidisciplinary design process. Within this chapter, a brief review of past methods for modelling and simulation of flexible aircraft is presented before the physics based modular approach is discussed. This is followed by details of the methods needed to integrate aerodynamics, structural dynamics and flight dynamics within a single simulation framework. Finally, the reader is presented with two test cases that demonstrate the use of such a framework in aircraft design. The Cranfield Accelerated Aeroplane Loads Model (CA^{2}LM) [5, 6] forms the basis of the discussion presented in this chapter.

2. Review of past methods

An extended version of the Collar's triangle shown in **Figure 1** highlights the physical phenomena that need to be integrated for accurate modelling and simulation of flexible aircraft. Traditionally the flight dynamics community has focused on the link between inertial dynamics and aerodynamics and it assumes structural dynamics to occur at far higher frequencies than those of rigid-body dynamics. The vice versa is true for the structural dynamics community who have mainly focused on specific loads cases for sizing airframe components. The development of aircraft such as the Boeing 747 [7], which was exceptionally large, and the Rockwell B-1 [8] with its flexible fuselage made it necessary for flight dynamics and structural dynamics to be integrated. The work done by Schmidt and Waszak [9] is an early example of such an integrated modelling approach carried out from a flight dynamicist's perspective. The approach retains the inertial components of the classical nonlinear six degree of freedom (6-DoF) equations [1, 2].

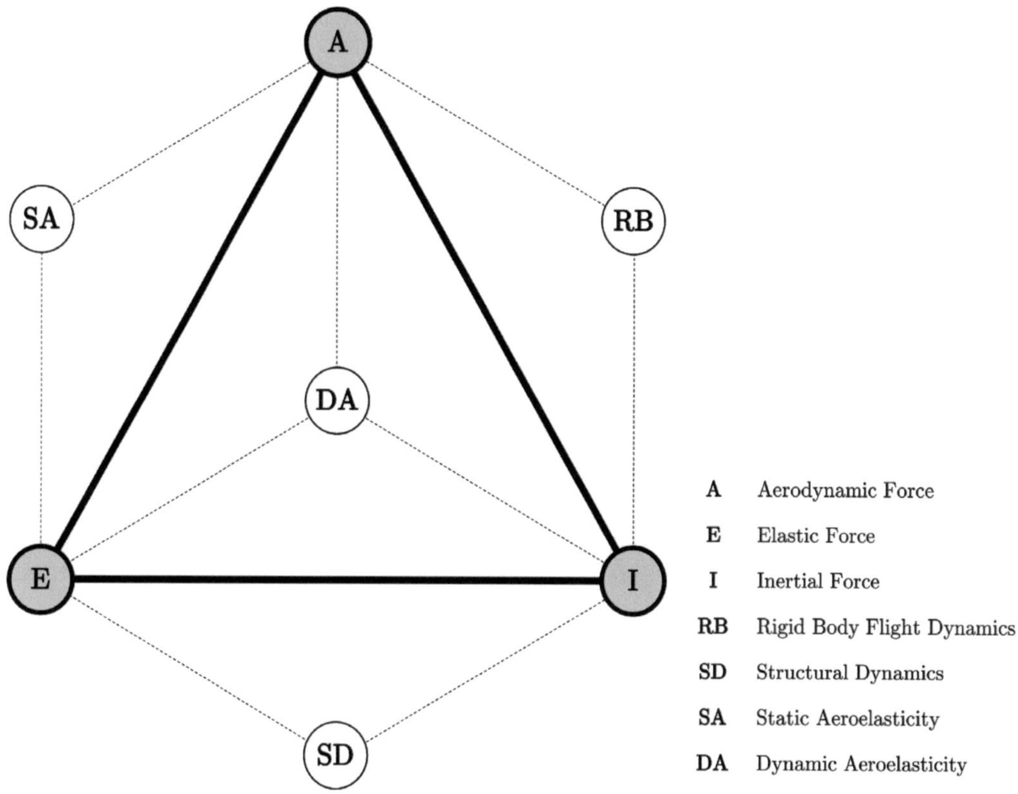

Figure 1. Extended Collar's triangle.

However, the aeroelastic effects are introduced by the addition of states related to each aeroelastic mode. Assuming that the free vibration modes are available, these make a set of orthogonal functions. The modal representation of the airframe is often obtained through the use of beam element models of the structure and the use of structural analysis software such as NASTRAN. Thus the airframe deformation $e(x,y,z,t)$ can be described in terms of the mode shape $\phi_i(x,y,z)$ and the general displacement coordinate $\eta_i(t)$, as follow:

$$e(x, y, z, t) = \sum_{i=1}^{\infty} \Phi_i(x, y, z)\eta_i(t) \tag{1}$$

The sum of the mode shapes is theoretically infinite but in practice, a finite number of mode shapes are selected in order to investigate the coupling of aeroelastic modes with rigid-body dynamics. The coupling between the rigid-body motion and elastic motion takes place through the forces and moments. The generic force and moment term can be described as function of the inputs (as in the general rigid equations of motion) and the generalised displacement η and its first derivative $\dot{\eta}$, as follow:

$$F = f(u, \alpha, \delta, \ldots, \eta, \dot{\eta}) \tag{2}$$

A new equation is then introduced to account for the elastic dynamics as:

$$\ddot{\eta} + \omega^2 \eta_i = \frac{Q_{\eta i}}{M_i} \tag{3}$$

where $Q_{\eta i}$ and M_i are the generalised force and mass terms, respectively. This formulation allows the application of stability analysis and flight control methods that have been developed based on traditional aircraft models.

Since the work done by Waszak and Schmidt, modelling frameworks of varying complexity have been developed both in industry and academia. Industrial frameworks are highly complex and aimed at supporting certification activities. These often couple Computational Fluid Dynamics (CFD) with Computational Structural Modelling (CSM) and result in processes that provide the desired insight, but at a very high computational cost [10–12]. Much research has been carried out to reduce the computational cost and the effort needed to integrate CFD solvers with CSM packages. However, more often the approach has depended on the specific technical challenge faced by the designer. For example, a few CFD-CSM simulations may be carried out to provide a means of validation for Reduced Order Models (ROMs). The various methods for aerodynamic and structural analysis are summarised in **Figure 2**.

Academic research has shown the capability to link aeroelasticity with flight control and develop novel approaches to aeroservoelastic analysis of highly flexible configurations [13–15].

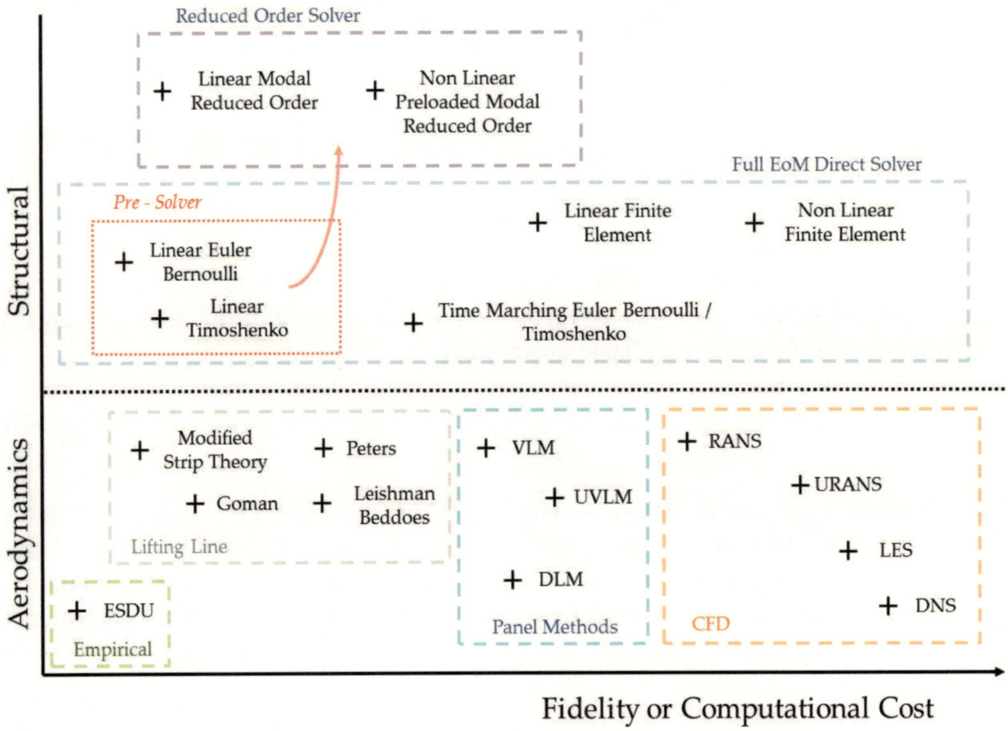

Figure 2. A non-exhaustive list of modelling methods ranked by complexity and fidelity.

Structural flexibility effects have been modelled through the implementation of a nonlinear structural dynamics formulation and aerodynamic contributions have been captured by means of an Unsteady Vortex Lattice Method (UVLM) code. Solving the geometrically-nonlinear beam equations in three different ways, Palacios et al. concluded that the intrinsic beam element model is more efficient regarding the computational time than the classical displacements and rotations based model. It has been shown that for certain geometries the intrinsic model required two times less operations per iteration due to simpler algorithms.

With regards to aerodynamic modelling Palacios et al. [14] showed that an indicial response based on the usual Pade approximation to Wagner's step response performs better at low reduced frequencies than the model based on a Glauert's expansion of the inflow velocity field. Three models—strip theory, strip theory with wingtip effects correction and UVLM—have been compared for different reduced frequencies and wingtip deflections. It has been shown that at low reduced frequency wingtip effects is of high importance both for low and high aspect ratio wings. However, for the case of increased reduced frequencies there has been no agreement of results for low aspect ratio wing. On the other hand, for high aspect ratio wing the agreement between the UVLM and the strip theory without wingtip correction has been shown. Such an agreement has been expected as increasing wing aspect ratio tends to reduce the 3D effect over the wing. The dynamic stall effects have not been modelled in the examples, nevertheless they may be of a great importance for a highly flexible wing. It is important to notice at this point that, if such a dynamic stall model is required by the user, empirical methods are much easier to implement within 2D strip theory than within the UVLM. Palacios and Cesnik [13] included aerofoil deformations in both the structural and the aerodynamic models: A Ritz (finite-section) expansion includes cross-sectional structural deformations, while a Glauert's expansion accounts for deformations of the aerofoil camber line. Integration of both expansions into a single methodology provides a simple alternative to more complex two-dimensional and three-dimensional models for preliminary active aeroelastic analysis of High Aspect Ratio Wings (HARW).

Although the approach adopted by Palacios is computationally cheaper than coupled CFD-CSM, real time simulation is still not possible. The need for real time simulation of flexible aircraft arises from the concern that low frequency aeroelastic modes can potentially couple with rigid-body modes such as the aircraft's short period pitch oscillation and result in poor handling qualities due to unwanted aircraft-pilot coupling [16]. Furthermore, novel concepts for future aircraft, such as those based on blended-wing-body configurations, need detailed stability and control analysis early in the design stage. A real time pilot-in-the-loop simulation environment is therefore needed to identify and solve stability and control problems. The development of such a simulation model requires a trade-off between model fidelity and computational cost.

3. Physics based modular approach

3.1. Aspects of physics based modelling

The case for developing physics based simulation models and the motivation to move away from the classical formulations that rely on stability and control derivatives stems from the need for flight dynamic insight at the early conceptual design of highly integrated concepts. For such concepts, a database of stability and control derivatives such as Heffley and Jewell [17] does not exist. Moreover, these concepts integrate numerous technologies, such as active folding wingtips for flight and loads control [18] for which empirical methods also do not exist. The modelling and simulation of airframe aerodynamics alone can be complex, but a further layer of complexity is added when considering flexible aircraft for which, the inertial, aerodynamic and structural models need to be coupled. Multiple calculation points, known as structural nodes and aerodynamic panels, must be defined around the airframe and used to capture local flow physics. The structural model must be coupled with the aerodynamics model so that aerodynamic forces and moments acting on the structure modify the effective shape of the aircraft. To complete such an aeroelastic coupling, the updated shape is used to compute the aerodynamic loading for the next iteration.

This additional layer of complexity and iteration process requires a clear definition of methods used when investigating aircraft flight dynamics. These can be broadly divided into two categories:

a. Low fidelity models used in particular for flight simulation and preliminary design studies. These allow for a rapid flight dynamic analysis and may allow parameters to be modified for identifying and quantifying possible optimised solutions.

b. High fidelity computationally expensive models which are used to consolidate the results obtained via low fidelity simulations and help in the investigation of specific problems where low fidelity simulation is not accurate.

For a given problem, multiple approaches can be adopted depending on the needs of the user or the key characteristics of the simulation framework. For example, the structural dynamics of the aircraft can be captured through the integration of a full Finite Element (FE) model with high fidelity, or with a simple beam, or 'stick' model. Within the latter method, multiple sublayers of complexity can be added depending on the mathematical formulation being used. A direct solving method, which is the most intuitive as it is based on discrete structural loads and nodes, will also be the most laborious and computationally heavy for a high number of structural elements. Alternatively, the modal approach restricted to frequency ranges of interest will be more efficient for linear deformations. In High Altitude Long Endurance (HALE) aircraft or HAR Wing concepts, structural nonlinearities can also become a physical phenomenon that must be captured by the model. Nonlinearities may be relevant only for specific modes and parts of the structure so that optimal solving methods can be identified as well.

Similarly, centre of gravity (CG) position and inertial terms will vary with structural flexibility and displacement. Therefore, acceptable or desired fidelity must be identified. For example, assuming a fixed CG and inertia can lead to significant simplifications in the EoM. However, this may be incorrect for HALE configurations where most of the mass lies in the flexible wing that undergoes large deformations.

Multiple methods to capture the aerodynamic loads acting on the aircraft have also been developed for different levels of fidelity; from simple lifting line theory, use of Engineering Science Data Unit (ESDU) to more complex UVLM and further to more expensive CFD based processes. The desired accuracy and performance can be optimised depending on the purpose of the framework. Dynamic stall models can also be added for a more accurate simulation of high angle of attack or flow detachment scenarios [19]. CFD simulations are at the higher fidelity end of the spectrum and can be used for construction of the aerodynamic databases [20].

3.2. Modular simulation

The objectives and scope of the problem being considered will undoubtedly dictate which mathematical formulation is selected. For instance, the aerodynamic forces can be calculated using either a Modified Strip Theory (MST) or a UVLM method [21] depending on the fidelity requirements and the available computational power. The structural deflection of the wing can be assumed either linear through an Euler-Bernoulli model or nonlinear with a Timoshenko model [22]. Various atmospheric disturbance models [23] are also implemented so that flight simulations with or without gusts and turbulence are possible for specific gust loads and flight control research. Flight control laws and actuation models of a variety of control surfaces can be used if the user wishes to investigate and develop optimal control or loads alleviation laws. The gravity and navigation model allows for trajectory and autopilot if required. Specialised hardware can be used to accelerate the model and reach real time performances suitable for pilot in the loop simulations at 50 Hz, paving the way for handling quality analysis of flexible aircraft concepts. So far a number of different modelling approaches towards flight dynamics modelling of flexible aircraft have been introduced. This section focuses on the possible problems and issues that emerge when integrating the various elements of such a framework and discusses the need for modularisation.

The basic components required for building a simulation framework are as follows:

1. A structural dynamics model that outputs airframe deformation. This should require forces and moments acting on the structure as inputs, and provide the corresponding displacements, velocities and accelerations as outputs.

2. An aerodynamic model that provides aerodynamic forces and moments as a function of the flight conditions, rigid-body attitudes and structural deformations.

3. An EoM block which uses the total forces and moments acting on the aircraft to compute the vehicle acceleration, velocity, attitude and position in the various reference frames. This will require a clear definition of aircraft mass properties.

4. Atmosphere model that outputs parameters such as Reynolds number required to calculate aerodynamic forces and moments.

5. A gravity model to compute the gravitational forces acting on the aircraft.

6. External atmospheric disturbances based on external velocity fields through which the aircraft is flying. This can be used for carrying out gust/turbulence simulations.

7. Control surface and flight control systems to simulate controlled flight.

Figure 3 illustrates the links between each of the modules and their relative dependencies.

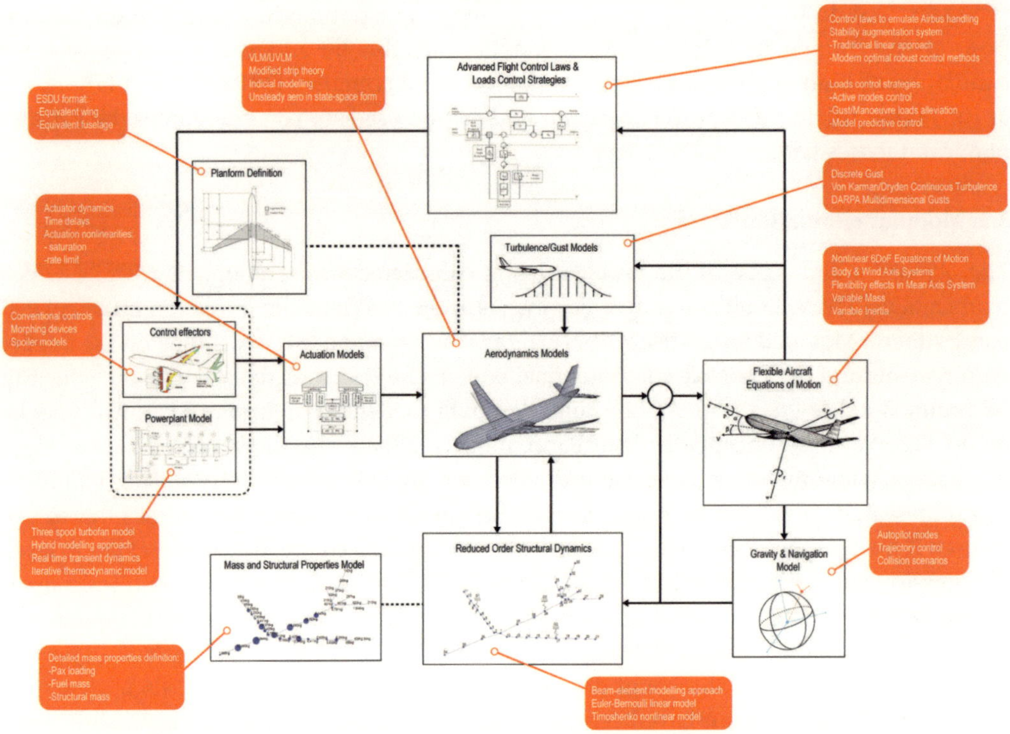

Figure 3. Links between each modules of the simulation framework.

Adopting a modular approach allows for a more versatile framework that can be used to study different configurations and scenarios. Moreover, it allows the adoption of multiple approaches to solve particular mathematical or physical problems. The overhead effort required to develop a modular framework, which primarily takes the form of software engineering, is justified by the end result. If carefully managed a versatile framework that allows solvers and models to be treated in a plug-and-play fashion is achievable. An example of a modular framework is given in **Figure 4**. The CA^{2}LM framework offers the user multiple options in most of the different mathematical models. The modular approach was considered at the early stages of framework development, and has allowed continuous development aiming for a versatile academic research tool.

4. Framework setup for CA^{2}LM

4.1. Wing aerodynamic modelling

There are numerous ways in which wing aerodynamics can be modelled for flexible wings, such as directly via CFD using RANS simulations or steady or unsteady VLM. However, given that there can be thousands of cases that need to be considered for flight loads, computationally cheap alternatives are needed. Within the CA^{2}LM framework, the aerodynamics module

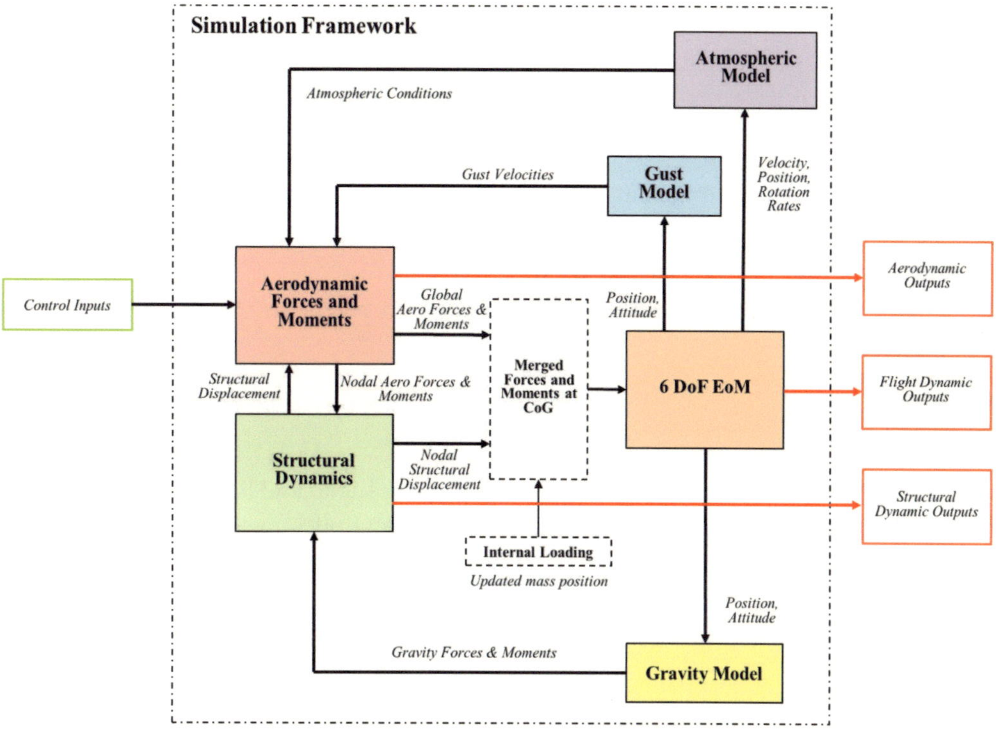

Figure 4. CA^{2}LM framework overall modular architecture.

contains the implementation of the MST based steady aerodynamics coupled with unsteady aerodynamic models [24].

To model the unsteady build-up of lift due to changes in angle of attack and airspeed, a state-space representation of the unsteady aerodynamics of the aerofoil has been implemented following the work done by Leishman and Nguyen [25]. This assumes an arbitrary motion of the aerofoil as combination of the indicial lift response and the superposition principle applying the well-known Duhamel's integral [26]. The following general two-pole approximation of the Wagner function has been adopted in CA^{2}LM:

$$\Phi(\lambda) \approx 1 - A_1 e^{-b_1 \lambda} - A_2 e^{-b_2 \lambda} \tag{4}$$

where $\lambda = 2V_t/c$ is the relative distance travelled by the aerofoil in terms of semi chords whilst A and b are the indicial response parameters that depend on the boundary conditions. Using the two-pole representation, Leishman and Nguyen developed the lift response to a change in angle of attack $\alpha(t)$ as follow:

$$\begin{bmatrix} \dot{x}_1 \\ \dot{x}_2 \end{bmatrix} = \frac{2V}{c} \begin{bmatrix} -b_1 & 0 \\ 0 & -b_2 \end{bmatrix} \begin{bmatrix} x_1 \\ x_2 \end{bmatrix} + \begin{bmatrix} 1 \\ 1 \end{bmatrix} \alpha(t) \tag{5}$$

and the output equation of the normal force coefficient is given by:

$$C_N(t) = 2\pi \frac{2V}{c} [A_1 b_1 A_2 b_2] \begin{bmatrix} x_1 \\ x_2 \end{bmatrix} \tag{6}$$

Coefficients A_i and b_i have been derived by Leishman in order to obtain the indicial response approximation for a two-dimensional subsonic flow [27]. However, since the Wagner indicial response cannot be applied to compressible flows, a correction introduced by Leishman and Beddoes [28], has been used including the Prandtl-Glauert coefficient $\beta = \sqrt{1 - M^2}$. The full equation of unsteady aerodynamics is then described as:

$$\phi(\tau) \approx 1 - 0.918e^{-0.366\beta^2} - 0.082e^{-0.102\beta^2} \tag{7}$$

Increasing the number of poles of the Wagner function allows a closer approximation to be obtained, but at the cost of an increased number of states.

In the CA^{2}LM framework the two-pole representation is used to find lift and pitching moment response with respect to a change in angle of attack α and pitch rate q for each section. The generic total normal force coefficient is given by [29, 30]:

$$C_N(t) = C_N^c(t) + C_{N\alpha}^{nc}(t) + C_{Nq}^{nc}(t) \tag{8}$$

where the superscripts c and nc represent the circulatory and non-circulatory terms respectively. Once aerodynamic characteristics are obtained at each aerodynamic node, the results are extended along the wingspan applying the method defined by DeYoung and Harper [31]. This approach considers the lift line and its trailing vortex as continuous. The circulation strength, however, can be discretized in as many control points as desired. In the CA^{2}LM framework the control points are assumed to be at the aerodynamic nodes. DeYoung and Harper stated that a number of seven control points is enough to correctly represent the span loading without any sharp discontinuities. As the lifting line is discretized in m nodes, the method allows the calculation of the aerodynamic coefficients as follows [29]:

$$cC_{lv} = \sum_{n=1}^{m} A_{vn} G_n \alpha_n, \qquad n = 1, 2, \ldots, m \tag{9}$$

where A_{vn} is the influence matrix which defines the effect of the circulation in the node v to the downwash at node n. The load coefficient G is dimensionless circulation and describes the strength of the circulation at any node n. When the aerodynamic forces and moments at each node are obtained, the loads are transposed from nodal-axis to body-axis and summed to give the overall lift, drag and moment acting on the aircraft structure.

Following the same methodology used for the calculation of the drag, the pitching moment is comprised of circulatory and non-circulatory term, described as follow:

$$C_M = C_{M_\alpha}^c + C_{M_q}^c + C_{M_\alpha}^{nc} + C_{M_q}^{nc} \tag{10}$$

The drag is instead modelled as the sum of the zero-lift drag coefficient, C_{D_0}, and the pressure drag coefficient, C_{D_p}. The unsteady drag force has been defined by Leishman as:

$$C_D = C_{D_0} + C_N \sin \alpha_e(t) - \eta_c C_c \cos \alpha_e(t) \tag{11}$$

where the effective angle of attack α_e is function of both the states and it is described as:

$$\alpha_e(t) = \beta^2 \frac{2V}{c}(A_1 b_1 x_1 + A_2 b_2 x_2) \tag{12}$$

and the chord force term is:

$$C_c(t) = \frac{2\pi}{\beta}\alpha_e^2(t) \tag{13}$$

As a real flow is unable to be fully attached in any real flow, the coefficient η_c is used to account for the properties of the real flow.

4.2. Structural modelling

Now all aerodynamic forces have to be applied to the structures of the aircraft. This is done in the structural dynamics modelling block.

Aerodynamic forces and moments, along with forces and moments due to gravity, are converted to modal forces F through modal transformation matrix $\boldsymbol{\Theta}_m^T$:

$$F_i = \boldsymbol{\Theta}_m^T \boldsymbol{F}_{aero} \tag{14}$$

The next step is to solve the following structural equation of motion:

$$\frac{F_i}{m_i} = \ddot{x}_i + 2\zeta\omega_n\dot{x}_i + \omega_n^2 x_i \tag{15}$$

where F_i represents the modal forces, m_i the modal masses, $\omega_{n,i}$ the modal natural frequencies, ζ the modal damping ratios, i is the modes number, $x_i, \dot{x}_i, \ddot{x}_i$ are the modal displacements, velocities and accelerations. To obtain the structural dynamics in modal form, the Normal Modes analysis solver SOL 103 from the NASTRAN finite element analysis program is used. Its output (modal masses, natural frequencies and modal transformation matrix) are used in the CA^2LM framework to calculate structural deflections. The displacements, velocities and accelerations of each structural node can then be obtained using the transformation matrix.

As these deflections, velocities and accelerations are applied to aerodynamic frame, the interpolation between structural and aerodynamic nodes is executed.

The first 12 structural modes are considered in the CA^2LM framework because the tool is designed to investigate interactions between aeroelasticity effects and flight dynamics phenomena that are typically at low frequencies. An illustration of an aircraft first four modes is given in **Figure 5**.

It is important to note that only small wingtip deflections (less than 10% of a wing semi-span) are modelled within CA^2LM framework as linearly varying beam properties are assumed. However, recent developments in highly flexible aircraft [32] have introduced wingtip

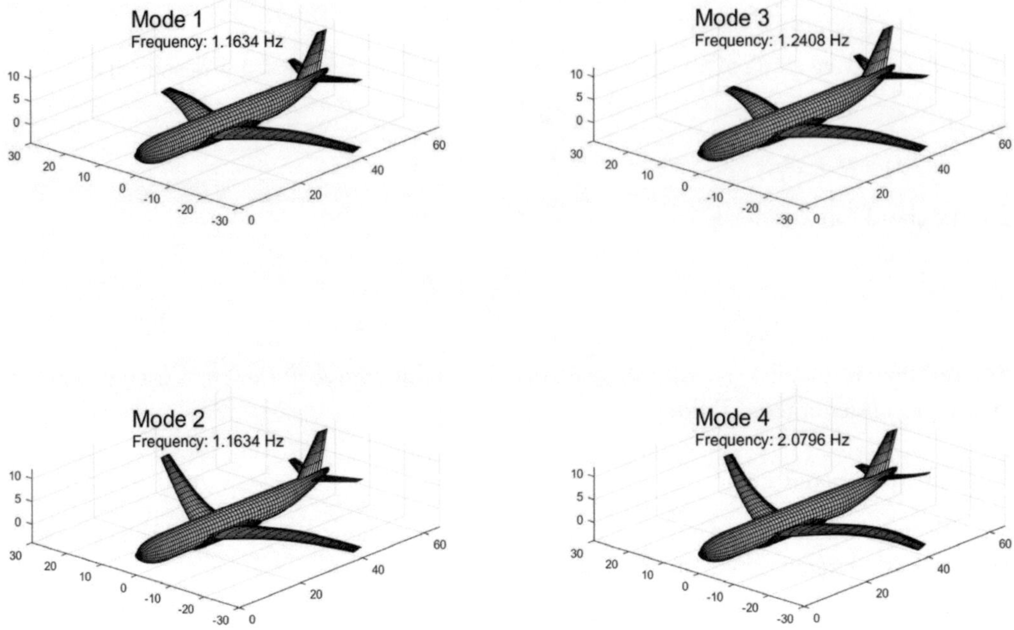

Figure 5. First four modes of the AX-1 aircraft implemented in CA^{2}LM.

deflections of more than 25% of a wing semi-span. To investigate the effects of such high structural deformations on flight dynamics, a structural dynamics model capable of capturing the nonlinear phenomena due to large deformations is needed.

4.3. Equations of motion

For large flexible aircraft, the centre of gravity (CG) position may vary significantly as a function of structural deformation. This is typically ignored in the classical EoM formulation for rigid aircraft [1, 2]. This issue together with continuously deforming aerodynamic and structural stations requires the careful definition of the axes systems for each module of the simulation framework. The selection of an appropriate axes system has been extensively discussed for many years [8, 33, 34]. Effectively there are two approaches that may be adopted: (1) use an arbitrary point on the aircraft also called the body axes centre (BAC) or, (2) adopt the mean axes system which has a floating point as the reference centre [35]. The latter has seen widespread application in research [9, 36] because its formulation minimises the coupling between rigid-body dynamics and aeroelastic modes. On the other hand, the axes system centre is often collocated with the CG which moves in phase with the flexible airframe, making the application of traditional flight dynamics analysis techniques more difficult. The idea of the mean axes system's inertial decoupling and complexity of its formulation has been questioned [34].

The CA^{2}LM framework uses a fixed BAC as a reference centre for its flight dynamic axis system. This allows the framework to be used in both flexible and rigid modes and more importantly, it allows the integration of classical flight dynamics post-processing tools.

http://dx.doi.org/10.5772/intechopen.71050

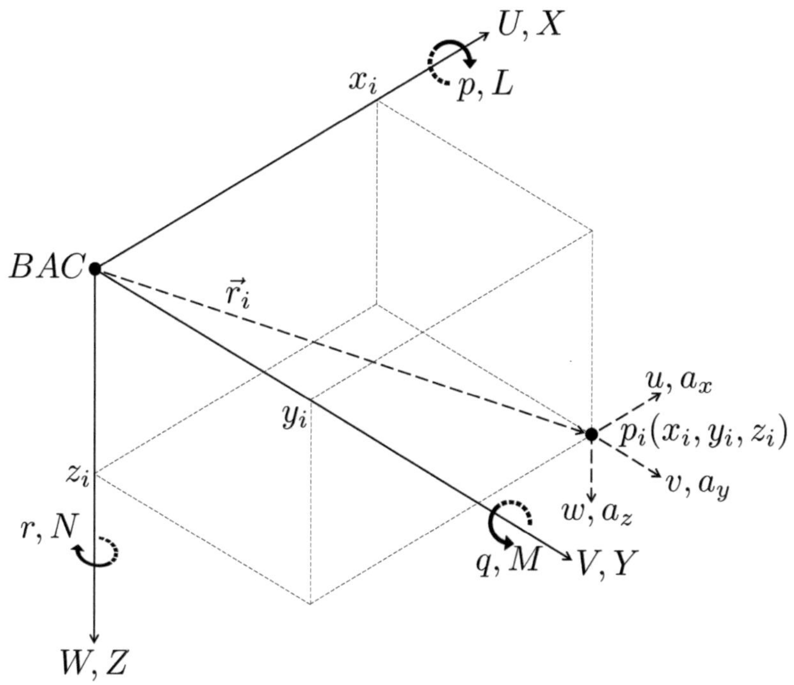

Figure 6. Motion of a body and its particle within the frame.

The derivation of the EoM begins by considering a fixed node which is located away from the BAC, as shown in **Figure 6**. The velocities of this point can be expressed as:

$$\begin{bmatrix} u \\ v \\ w \end{bmatrix} = \begin{bmatrix} U+\dot{x} \\ V+\dot{y} \\ W+\dot{z} \end{bmatrix} - \begin{bmatrix} x \\ y \\ z \end{bmatrix} \times \begin{bmatrix} p \\ q \\ r \end{bmatrix} \tag{16}$$

And therefore, the following accelerations can be obtained:

$$\begin{bmatrix} a_x \\ a_y \\ a_z \end{bmatrix} = \begin{bmatrix} \dot{u} \\ \dot{v} \\ \dot{w} \end{bmatrix} - \begin{bmatrix} u \\ v \\ w \end{bmatrix} \times \begin{bmatrix} p \\ q \\ r \end{bmatrix} \tag{17}$$

The velocities U, V and W express the motion of the BAC, while x, y and z express the position of the node. The angular rates p, q and r represent the angular velocities of the overall aircraft. Merging both equations gives following accelerations expressions:

$$a_x = \dot{U} - rV + qW - x(q^2 + r^2) + y(pq - \dot{r}) + z(pr + \dot{q}) + \ddot{x} - 2r\dot{y} + 2q\dot{z} \tag{18}$$

$$a_y = \dot{V} - pW + rU + x(pq + \dot{r}) - y(p^2 + r^2) + z(qr - \dot{p}) + \ddot{y} - 2p\dot{z} + 2r\dot{x} \tag{19}$$

$$a_z = \dot{W} - qU + pV + x(pr - \dot{q}) + y(qr + \dot{p}) - z(p^2 + q^2) + \ddot{z} - 2q\dot{x} + 2p\dot{y} \tag{20}$$

Now applying Newton's second law with a nodal mass of δm the EoM can be obtained as follows:

$$\begin{bmatrix} X \\ Y \\ Z \end{bmatrix} = \sum_{i=1}^{N} \delta m_i \begin{pmatrix} a_x \\ a_y \\ a_z \end{pmatrix} = \overbrace{\sum_{i=1}^{N} \delta m_i \dot{v}_0 + \sum_{i=1}^{N} \delta m_i \omega \times v_0}^{\textit{Rigid body dynamics force}}$$
$$+ \overbrace{\overbrace{\sum_{i=1}^{N} \delta m_i \omega \times (\omega \times r_i)}^{\textit{Centrifugal force}} + \overbrace{\sum_{i=1}^{N} \delta m_i \dot{\omega} \times r_i}^{\textit{Euler force}}}^{\textit{Axes reference point offset}} \tag{21}$$
$$+ \underbrace{\underbrace{\sum_{i=1}^{N} \delta m_i a_{rel,i}}_{\textit{Inertial force}} + \underbrace{2 \sum_{i=1}^{N} \delta m_i \omega \times v_{rel,i}}_{\textit{Coriolis force}}}_{\textit{Flexibility effects}}$$

$$\begin{bmatrix} L \\ M \\ N \end{bmatrix} = \overbrace{I\dot{\omega} + \omega \times (I\omega)}^{\textit{Rigid body dynamics}} + \overbrace{\sum_{i=1}^{N} \delta m_i r_i \times \dot{v}_0 + \sum_{i=1}^{N} \delta m_i r_i \times (\omega \times v_0)}^{\textit{Axes reference point offset}}$$
$$+ \underbrace{\dot{I}\omega + \omega \times \sum_{i=1}^{N} \delta m_i (r_i \times v_{rel,i}) + \sum_{i=1}^{N} \delta m_i r_i \times a_{rel,i}}_{\textit{Flexibility effects}} \tag{22}$$

The forces and moments on the left hand side of the above equations are the sum of the forces and moments obtained from the structural dynamics, aerodynamics and gravitational modules.

4.4. Aeroelastic coupling and equations of motion integration

The previous sections have shown that each module within the simulation framework requires the definition of its own axis system and a separate means of modelling the aircraft, whether it is through a set of structural nodes or aerodynamic panels. This presents two issues that must be addressed before scenarios can be simulated: (1) node and panel distributions and densities need to be optimised based on the scope of the research and, (2) the structural nodes must be linked to aerodynamic nodes.

As seen in the previous section, the structural loads calculations rely on a set of structural nodes. Displacements, velocities and accelerations of each node are calculated in all 6 degrees of freedom.[1]

---

[1]It is possible to constrain specific degrees of freedom to reduce model complexity after a comparison study with the 6 DoF model. For stiff wings, structural rotation around the vertical axis can be neglected for example.

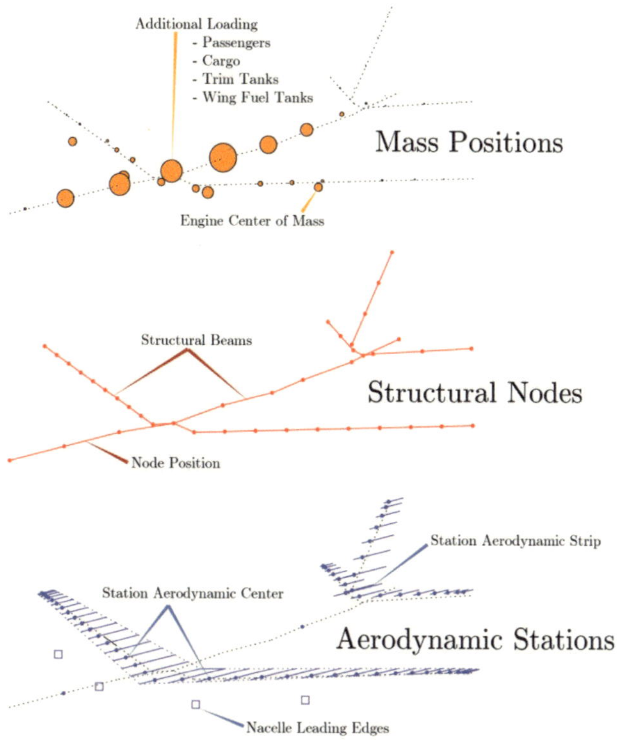

Figure 7. Illustration of the different mass, structural node and aerodynamic station positions for the AX-1 aircraft.

Appropriate balance between accuracy and computational cost must be obtained using a convergence study to identify the optimal number of structural nodes and aerodynamic panels or strips. This number can vary with aircraft configuration and the type of flight dynamics being considered. However, the number of structural nodes may be different from the optimal number of aerodynamic stations. A modular simulation environment such as CA^2LM allows the definition of different numbers of aerodynamic strips and structural nodes. The aerodynamic forces and moments calculated at the aerodynamic stations must then be transferred to the structural set of nodes using various interpolation methods. Similarly the structural displacements, velocities and accelerations calculated from the structural model must be transferred to the aerodynamic stations in order to calculate the local forces and moments with structural flexibility. An example of this coupling can be found in **Figure 7** where both the structural node and aerodynamic station layout is illustrated for an example aircraft.

The EoM rely on the aircraft total forces and moments, acting around the centre of gravity of the vehicle. Therefore, the updated CG position due to structural deformation must be used to calculate the new global set of moments acting on the aircraft. Aerodynamic loading calculated at each aerodynamic station is merged and calculated at the temporary CG position. Only then can the coupling between the aerodynamic and structural block be made with the EoM.

The output of the EoM such as aircraft position, attitude and velocity can then be used by conventional atmospheric models to compute the dynamic pressure and other aerodynamic

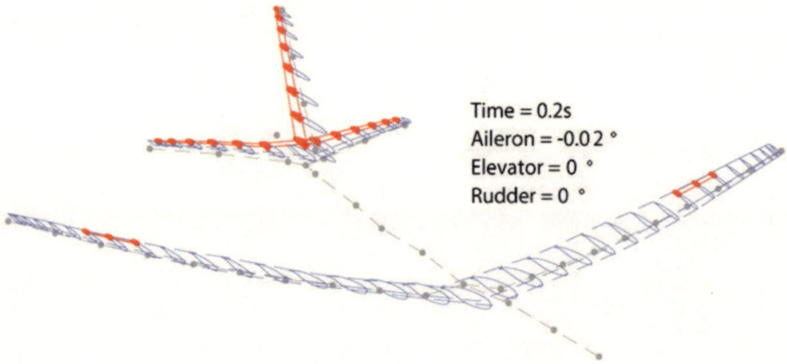

Figure 8. Aircraft flexible structure overlaid with aerodynamic profiles and control surfaces for pilot input visualisation.

parameters used by the aerodynamic model, closing the main calculation loop. Similarly, the adequate gravity contribution can be computed with position (or altitude) and applied to the structural model.

Appropriate inputs, usually on aircraft control surface and thrust, should be linked to the model in the correct format. Control surface dynamics can be implemented for higher fidelity.

As each module is included in the simulation framework, correct integration testing must be conducted to verify that each modules are behaving as expected. Therefore, as the complexity of the framework increases, thorough testing also requires more effort. It can also be really helpful to have visual aids and illustrations of the simulation. For example, an illustration of aerodynamic station and structural node positions updated with structural flexibility at each time step can be found in **Figure 8** and is very useful to visualise the modelled aircraft.

5. Framework test cases

5.1. Multidimensional discrete gust loads simulation

The aim of this test case is to demonstrate the use of simulation frameworks such as CA^{2}LM for assessing the impact of multidimensional discrete gust modelling on conventional gust loads practices seen in industry. The prediction and control of aircraft gust loads is a key step in aircraft design development and certification. The methodology to model realistic discrete and continuous atmospheric disturbances has been derived based on many years of flight testing and operational data [37]. Hoblit [23] covers a concise but thorough overview of the historical development of gust and turbulence modelling in whereas a detailed discussion of current industry practices can be found in [35]. However, the methods to date simplify the process of calculating gust loads by neglecting spanwise variations in the gust/turbulence fields. This case study demonstrates the application of the CA^{2}LM framework for studying gust profiles that have spanwise variations. Atmospheric disturbances are usually added through the use of velocity fields. For each aerodynamic station, the wind or gust velocities can be added to the rigid-body translation, rotation and elastic structural dynamics in a local nodal axis system to compute local changes to angle of attack and flow velocities. If gusts are defined as a velocity field, the gust model should also use the aerodynamic station layout and aircraft attitude to apply a penetration effect.

With the development of HALE UAV aircraft, the lack of spanwise non-uniform velocity distributions was identified as critical both for realistic and theoretical modelling purposes. The gust profiles specified in certification requirements [37, 38] implicitly assume that a uniform velocity distribution causes the highest internal loads and therefore, are the only cases that need to be investigated. Therefore, Defense Advanced Research Projects Agency (DARPA) focused on the derivation of a modified discrete gust model to account for the extra dimensional term and led to the expression of the discrete gust velocity V_{dg} to be defined by:

$$V_{dg}(x_d, y_d) = V_{do} f_x(x_d, H_x) f_y(y_d, H_y) \tag{23}$$

where:

$$f_x = \frac{1}{2}\left(1 - \cos\left(\frac{\pi x_d}{H_x}\right)\right) \tag{24}$$

and f_y is the corresponding sinusoidal function. V_{do} is the gust intensity, H_x and H_y are the longitudinal and lateral gust gradients respectively and x_d and y_d are the longitudinal and lateral positions of the interest point in the discrete gust reference frame. Specifications to the range of both gust gradients can be made using similar hypothesis as before, ranging from 9 to 107 m.[2] An illustration of the multidimensional discrete gust velocity field is given in **Figures 9** and **10**.

This type of model was implemented as a feature within the CA^{2}LM framework and applied to a conventional long range flexible aircraft configuration known as the AX-1. A study investigating the impact of such an approach to gust loads prediction for conventional aircraft was then carried out [39] using a sinusoidal lateral distribution as follows:

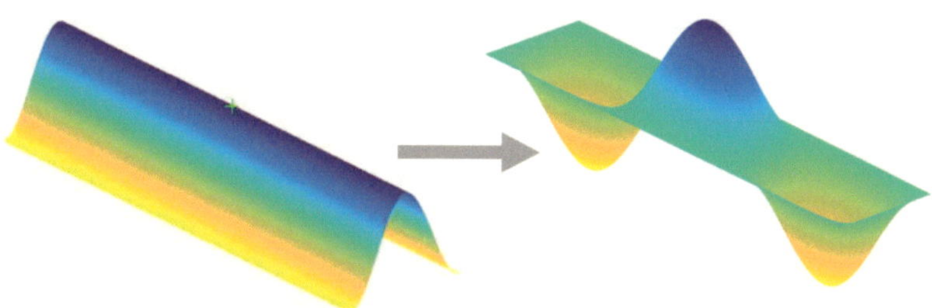

Figure 9. From a 1D to 2D discrete gust definition using coupled sinusoidal variation functions.

[2] In fact, it is necessary to push the higher end of the gradient spectrum so as to reach a minimum of 12.5 times the maximum aerodynamic chord of the vehicle and/or reach the peak maximum of the evaluated quantity with respect to the various conditions.

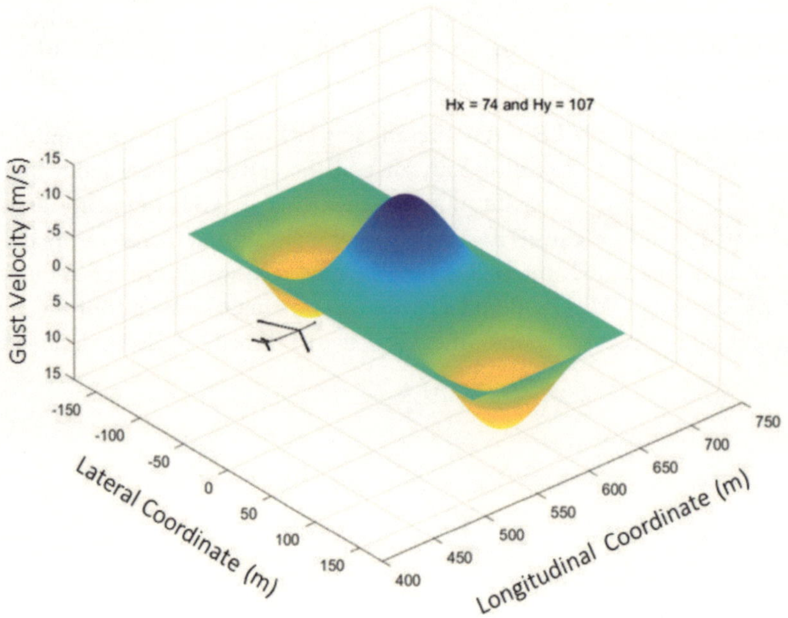

Figure 10. Visual display of the discrete gust velocity field for a given set of gust gradients used in the loads prediction loop.

$$f_y = \cos\left(\frac{\pi y_d}{H_y}\right) \tag{25}$$

A sufficiently large number of realistic flight points compatible with the framework and implemented aircraft were used for this study. A number of gust gradients were used to allow a comparison between the conventional spanwise uniform velocity field and the multi-dimensional model of interest with enough fidelity. All simulations were made in an open loop system, where no correction to aircraft attitude is made. Two different approaches were used to scale the maximum gust intensity, keeping the core hypothesis of the certification requirements. This is justified by the very nature of the derivation of the original model, based on flight testing and loads data and not actual mapping of the gust velocity fields.

In both cases, the use of a multidimensional model led to lower gust structural wing root loads and vertical loads for an equivalent longitudinal gust gradient, as illustrated in **Figure 11**. In one case of velocity tuning methodology, some local loads extrema were higher than with the conventional model, possibly leading to higher occurrence numbers of specific load values. This also came to a cost in computation time, increasing by an order varying with H_y discretisation size the number of simulations required for a complete gust loads loop process.

Overall, these results were to be expected with the chosen spanwise distribution. Maximal gust intensity was centred on the fuselage in this study. But these results can vary quite dramatically with the selected f_y distribution. If focused on matching the vertical load factor whilst keeping wingtip loads to the highest, this could lead to:

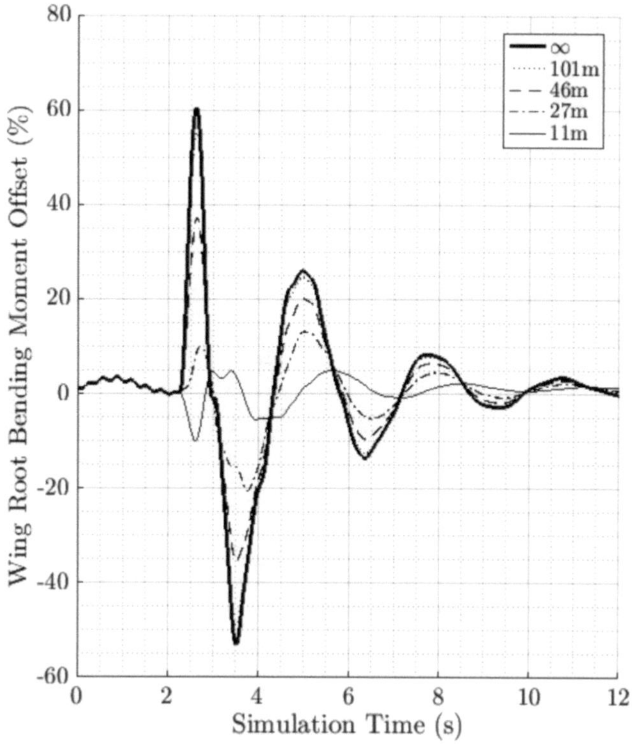

Figure 11. Time histories of wing root bending offset relative to trim for a given H_x and various H_y gradients.

- A 'realistic' model of the gust velocity field compliant with the historical development of the methodology based on vertical load and angle of attack data recordings.
- Higher wing root structural loads due to increased wingtip loading.

5.2. Aileron failure simulations

A control surface failure scenario is one of many failure cases that need to be considered for flight loads evaluation. Here the CA^{2}LM framework is used for simulating a soft aileron failure where the port aileron undergoes an actuation failure and is forced to undergo a 15° amplitude limit cycle oscillation (LCO) whilst starboard aileron remains in the original trim setting. The dynamics of the aileron actuators are modelled through the transfer function:

$$\delta_a(s) = \frac{-1.77s + 399}{s^2 + 48.2s + 399} \tag{26}$$

The main results obtained from the simulation of the AX-1 model are shown in **Figure 12**. The port aileron moves under a limit cycle oscillation at a constant frequency of 1.16 Hz, which corresponds to the first wing structural bending mode. The amplitude of this oscillation is set to $\pm 15^{\circ}$.

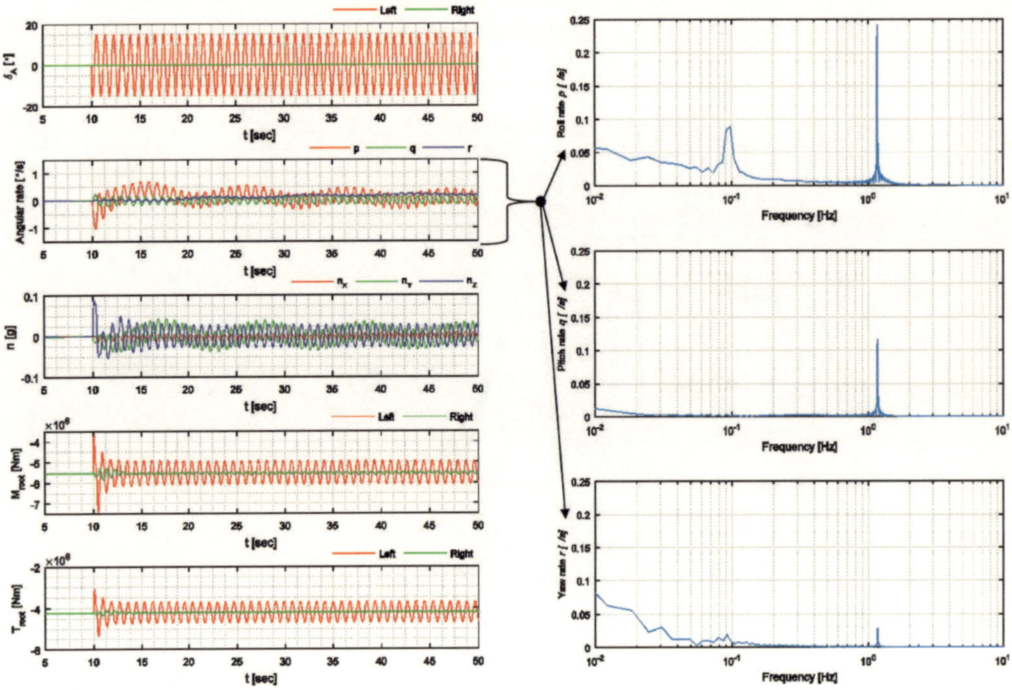

Figure 12. Example of AX-1 aileron cycle oscillation failure simulation results.

The frequency content of the roll rate p and yaw rate r signals show that the failure has excited a low frequency lateral-directional mode corresponding to periods of $T_p = 10.24$ s and $T_r = 10.92$ s in roll and yaw respectively. These correspond to the usual frequencies of the aircraft's Dutch roll mode. The highest peaks, just above 1 Hz, are the direct result of the simulated aileron forcing function. The load factor (n) only exhibits large transients when the aileron failure is initiated.

Figure 13 shows the frequency content of the wing root bending moment M_{root} at different aileron excitation frequencies. At a frequency of 1.245 Hz, slightly higher than the frequency of the first structural mode of the wing (1.1634 Hz), the first aeroelastic mode appears and a resulting resonance is observed. Upon magnification (bottom right subfigure) another two peaks can be observed at 2.5 and 3 Hz. These correspond to aeroelastic modes associated with the 5th and 11th aircraft structural modes. At the frequency of 0.9 Hz, M_{root} is higher than at the frequency of 1.1 Hz, which can be explained by the fact that the forcing function frequency is getting closer to rigid-body frequencies.

Simulations like this provide the insight loads engineers and flight control engineers need for exploring scenarios where a novel solution could be tested and design improvements can be made. Simulation frameworks such as CA^2LM provide a rapid simulation capability needed especially at low technology readiness levels, where engineers and designers are interested in the impact of novel technologies such as folding wingtips, possible aircraft-pilot coupling scenarios [40] and flight loads during collision avoidance [6].

http://dx.doi.org/10.5772/intechopen.71050

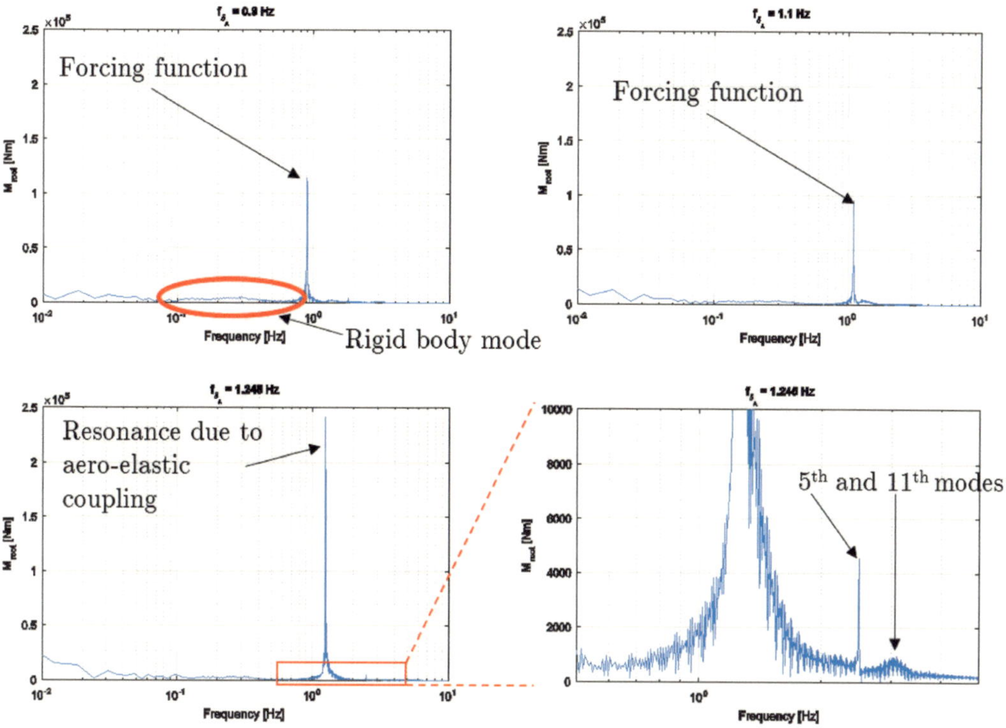

Figure 13. Wing root bending moment frequency spectrum for different aileron excitation.

6. Conclusions

Technologically innovative and highly integrated concepts are being considered in response to increasing aircraft efficiency and reducing the environmental impact of aviation. The development of these concepts has highlighted the need for modular low fidelity aircraft simulation frameworks at the conceptual design stage that are capable of predicting the flight dynamics, flight loads and aeroservoelastic characteristics. This chapter has presented the key aspects of developing such a framework and the need for a modular physics based approach. This approach requires a careful integration of aerodynamic models with models for structural dynamics and then both need to be coupled with the flight dynamic equations of motion. It has been shown that the aerodynamic representation must include a combination of unsteady and steady aerodynamic models implemented through aerodynamic panels. These panels need to then be linked to the aircraft structure which is typically implemented as a series of nodes and beams. The coupled aero-structural model then needs to provide forces and moments to the equations of motion. The details of developing such a simulation framework has been presented in this chapter and the utility of such a tool is illustrated through two test cases. The first case focuses on aircraft response to a gust that has a spanwise varying profile. The second investigates aircraft dynamics during control surface failure scenarios. The Cranfield Accelerated Aeroplane Loads Model (CA^{2}LM) forms the basis of the presented discussion.

Author details

Gaétan Dussart, Vilius Portapas, Alessandro Pontillo and Mudassir Lone*

*Address all correspondence to: m.m.lone@cranfield.ac.uk

Cranfield University, Cranfield, United Kingdom

References

[1] Cook MV. Flight Dynamic Principles: A Linear Systems Approach to Aircraft Stability and Control. 3rd ed. Oxford, United Kingdom: Butterworth-Heinemann Elsevier Ltd; 2012. 608 p

[2] Stengel RF. Flight Dynamics. Princeton, United States: Princeton University Press; 2015. 864 p

[3] Klein V, Morelli E. Aircraft System Identification: Theory and Practice. Reston, VA: AIAA; 2006. 484 p

[4] Jategaonkar RV. Flight Vehicle System Identification: A Time Domain Methodology. Progress in Astronautics and Aeronautics. Reston, USA: AIAA; 2006. 534 p

[5] Portapas V, Cooke A, Lone M. Modelling framework for flight dynamics of flexible aircraft. Aviation. 2016;**20**(4):173-82

[6] Lone MM, Lai CK, Cooke A, Whidborne JF. Framework for flight loads analysis of trajectory-based manoeuvres with pilot models. Journal of Aircraft. 2014;**51**(2):637-50

[7] Hanke CR. The simulation of large jet transport aircraft. NASA CR-1756. Washington, DC, United States; 1971

[8] Waszak MR, Buttril CS, Schmidt DK. Modeling and Model Simplification of Aeroelastic Vehicles: An Overview. Langley: NASA; 1992

[9] Waszak MR, Schmidt DK. Flight dynamics of aeroelastic vehicles. Journal of Aircraft. 1988;**25**(6):563-71

[10] Cooper JE, Gaitonde A, Dorian Jones MHL, Sartor P, Lemmens Y. Aircraft loads prediction using enhanced simulation. In: 15th Dynamics Specialists Conference, AIAA SciTech Forum. San Diego, CA; 2016. pp. 1-14

[11] Lindhorst K, Haupt MC, Horst P. Efficient surrogate modelling of nonlinear aerodynamics in aerostructural coupling schemes. AIAA Journal. 2014 Sep 26;**52**(9):1952-66

[12] Wang F, Huo S, Qiao S, Zhang J, Yue Z. An effective computer modelling approach to the study of aeroelastic characteristics of an aircraft composite wing with high aspect ratio. Mathematical and Computer Modelling of Dynamical Systems. 2015 Jan 2;**21**(1):58-76

[13] Palacios R, Cesnik CES. Low-speed aeroelastic modeling of very flexible slender wings with deformable airfoils. In: 49th AIAA/ASME/ASCE/AHS/ASC Structures, Structural Dynamics, and Materials Conference. Schaumburg, Illinois; 2008. pp. 1-19

[14] Palacios R, Murua J, Cook R. Structural and aerodynamic models in nonlinear flight dynamics of very flexible aircraft. AIAA Journal. 2010;**48**(11):2648-59

[15] Simpson RJ, Palacios R, Goulart PJ. Integrated flight dynamics and aeroelasticity of flexible aircraft with application to swept flying wings. In: 56th AIAA/ASCE/AHS/ASC Structures, Structural Dynamics, and Materials Conference. Kissimmee, Florida: AIAA; 2015. pp. 1-17

[16] McRuer T, Droste C, Hess R, LeMaster D, Mathhews S, McDonnel J, et al. Aviation safety and pilot control: Understanding and preventing unfavorable pilot-vehicle interactions. Washington, D.C., United States: National Academies Press; 1997. 220 p

[17] Heffley RK, Jewell WF. Aircraft Handling Qualities Data. Washington, DC, United States: NASA CR-2144; 1972 Dec

[18] Castrichini A, Hodigere Siddaramaiah V, Calderon DE, Cooper JE, Wilson T, Lemmens Y. Nonlinear folding wing tips for gust loads alleviation. Journal of Aircraft. 2016 Sep 17;**53**(5): 1391-9

[19] Goman M., Khrabrov A. State-space representation of aerodynamic characteristics of an aircraft at high angles of attack. Journal of Aircraft. 1994 Sep 22;**31**(5):1109-15

[20] Da Ronch A, Vallespin D, Ghoreyshi M, Badcock K. Evaluation of dynamic derivatives using computational fluid dynamics. AIAA Journal. 2012;**50**(2):470-84

[21] Katz J, Plotkin A. Low-Speed Aerodynamics. 2nd ed. Cambridge, United Kingdom: Cambridge University Press; 2001. 613 p

[22] Timoshenko S. History of Strength of Materials: With a Brief Account of the History of Theory of Elasticity and Theory of Structures. North Chelmsford, MA: Courier Corporation; 1953. 452 p

[23] Hoblit F. Gust Loads on Aircraft: Concepts and Applications. Washington, DC: AIAA; 1988

[24] Andrews SP. Modelling and simulation of flexible aircraft: Handling qualities and active load control [Thesis]. Cranfield University, United Kingdom; 2011

[25] Leishman JG, Nguyen KQ. State-space representation of unsteady airfoil behavior. AIAA Journal. 1990;**28**(5):836-44

[26] Bisplinghoff R, Ashley H, Halfman R. Aeroelasticity. North Chelmsford, Massachusetts: Courier Corporation; 2013. 880 p

[27] Leishman JG. Unsteady lift of a flapped airfoil by indicial concepts. Journal of Aircraft. 1994;**31**(2):288-97

[28] Leishman J, Beddoes T. A generalised model for airfoil unsteady aerodynamic behaviour and dynamic stall using the indicial method. In: 42nd Annual Forum of the American Helicopter Society. Washington, DC; 1986. pp. 243-65

[29] Andrews S, Cooke A. An aeroelastic flexible wing model for aircraft simulation. In: 48th AIAA Aerospace Sciences Meeting Including the New Horizons Forum and Aerospace Exposition. Orlando, Florida; 2010

[30] Leishman JG. Validation of approximate indicial aerodynamic functions for two-dimensional subsonic flow. Journal of Aircraft. 1988;**25**(10):914-22

[31] DeYoung J, Harper CW. Theoretical symmetric span loading at subsonic speeds for wings having arbitrary plan form. NACA Report No. 921. Langley, Virginia: NACA; 1948

[32] Mayuresh JP, Dewey HH. Flight dynamics of highly flexible flying wings. Journal of Aircraft. 2006;**43**(6):1790-9

[33] Guimarães Neto AB, Silva RGA, Paglione P, Silvestre FJ. Formulation of the flight dynamics of flexible aircraft using general body axes. AIAA Journal. 2016 Nov;**54**(11):3516-34

[34] Meirovitch L, Tuzcu I. The lure of the mean axes. Journal of Applied Mechanics. 2007 May 1;**74**(3):497

[35] Wright JR, Cooper JE. Introduction to Aircraft Aeroelasticity and Loads. Hoboken, New Jersey: John Wiley & Sons; 2008. 488 p

[36] Milne R. Dynamics of the deformable aeroplane. Aeronautical Research Council, Reports and Memoranda. London, United Kingdom; 1964

[37] Airworthiness Standards: Transport Category Planes. Washington, DC, United States: Federal Aviation Regulations; 2013

[38] Certification Specifications and Acceptable Means of Compliance for Large Aeroplanes CS-25. Cologne, Germany: European Aviation Safety Agency; 2012. p. 885

[39] Dussart G, Lone M, Guo S. Multidimensional discrete gust loads of a large civil flexible aircraft. In: AIAA Atmospheric Flight Mechanics Conference, AIAA SciTech Forum. Grapevine, Texas; 2017. pp. 2017-0016

[40] Lone M. Pilot modelling for airframe loads analysis [Thesis]. Cranfield University, United Kingdom; 2013

Chapter 4

Aerodynamic Characteristics and Longitudinal Stability of Tube Launched Tandem-Scheme UAV

Illia S. Kryvokhatko and Oleksandr M. Masko

Additional information is available at the end of the chapter

http://dx.doi.org/10.5772/intechopen.70228

Abstract

Tube launched unmanned aerial vehicles (UAV) are often implemented with aerodynamic scheme with forward and rear wings (so-called tandem-scheme). Specificity of such UAV is that immediately after launch, they have a flight path in which wings are turned from position along the fuselage to flight position in which sweep angles are about zero. UAV aerodynamic characteristics for different wing rotation angles were researched by computational fluid dynamics (CFD) methods (Ansys 16 software). Quantitative results prove that UAV is unstable with wings rotation angles up to 60° because rear wings produce lift ahead of center of gravity. Therefore, low time of wings unfolding is required. For high angles of wings rotation (low sweep angles), UAV model is stable in a wide range of angles of attack. Local aerodynamic defects were found in the area of the rotation units of both wings. Longitudinal vortex along the left side of fuselage was observed, but it does not result in significant roll moment. Further research might include UAV dynamics modelling based on calculated aerodynamics characteristics or flight tests.

Keywords: tandem-scheme UAV, tube launch, aerodynamic performance, flow separation, longitudinal static stability, computational fluid dynamics, Ansys software

1. Introduction

Tube launched unmanned aerial vehicles (UAV) are often implemented with so-called tandem-scheme, i.e., an aerodynamic scheme with forward and rear wings (**Figure 1**). It allows decreasing of the UAV size in folded state inside the tube, e.g., wingspan might be 30–40% <1 for a conventional aerodynamic scheme. So this aerodynamic scheme has become quite widespread in the twenty-first century [1–3].

Figure 1. Tube launched UAVs: "Switchblade," "Sokil-2," "Trident" (in folded state).

Specificity of such UAVs is that immediately after launch, they have a flight path in which wings are turned from the position along the fuselage to flight position with sweep angles equal to about zero [4].

The **goal** of this work is the determination of the aerodynamic characteristics of tube launched tandem-scheme UAV after its start, at the time of wings' unfolding.

2. Object and methods

Analytic methods are not applicable for such tasks because for low-wing rotation angles, there is a high sweep angle and, therefore, a low aspect ratio (**Figure 2**). For such a type of geometry, flat section hypothesis is not correct, and more complicated methods must be used.

In view of the experimental research expensiveness, it is reasonable to determinate aerodynamic characteristics with the help of computational fluid dynamics (CFD) methods. Ansys 16.0 software was used that allows proper determining of pressure distribution throughout

Figure 2. UAV model geometry for wings' rotation angles equal to 30° (sweep angles equal 60°).

UAV surface and, therefore, appropriate lift and moment coefficients (**Figure 3**). Unstructured meshes were built, and the results independence of cells number was proved for every case. One wing chord equals 110 mm and maximal size of mesh elements are following;

- inlet and far-field—500 mm;
- pressure-outlet—400 mm;
- wing, fuselage—10 mm;
- fins—5 mm;
- fuselage nose and wingtips—2 mm;
- leading edges curves of wings and fins—0.5 mm; and
- trailing edges curves of wings and fins—1 mm.

Total height of prism layer is 2 mm which approximately corresponds to maximal boundary layer thickness at wings' trailing edges. There was only one layer of prismatic elements, i.e., mesh was not sufficient to present shear stress and friction drag values in the boundary layer, but as lift and moment coefficients were the priority of this research, this approach that saves a great amount of time is appropriate.

Domain size equals to about 18 spans of rear wing as a larger model dimension. Mesh created in ICEM CFD software was converted to a polyhedral type that decreases calculation time greatly. Final mesh consists of 1.0–1.5 millions of polyhedral cells depending on the wings' rotation angles value.

Use of Menter's turbulence model (k-ω SST) is recommended for aerodynamic characteristics' determination at low and high angles of attack when separation occurs [5, 6]. Only a steady case was considered. A pressure-based solver was used with a coupled scheme and a second-order upwind for all available parameters. Computation model includes

Figure 3. UAV surface mesh with wings' rotation angles equal to 60°.

energy equations as well. Air density was defined as ideal gas, and viscosity was set according to the Sutherland model. Pseudo-transient factors and all unmentioned parameters were set on default. All UAV surfaces had roughness equal to 0.05 mm according to wind tunnel model demands. Gauge pressure of 101,325 Pa, velocity of 25 m/s, and temperature of 288 K were set. Calculation process was finished after all residuals had reached 10^{-4}. For high angles of attack, this threshold was not always achievable, so iteration process was stopped after residuals had become minimal and force coefficients had become about constant.

The object has typical geometric parameters of tube launched UAV with equal chords of all wings and the rear wing span slightly bigger than the forward wing span. In the unfolded state, distance between leading edges of forward and rear wings (stagger) was equal to about 4 chords of one wing. Vertical gap between forward and rear wings from one side is about 70% of wing chord. Gap between left and right wings equals 10% of wing chord, i.e., slightly bigger than airfoil thickness. Wings' rotation angles of 15, 30, 45, 60, 75, and 90° were considered (equal for forward and rear wings). So wings' rotation angle of 15° means forward wing swept angle of 75° and rear wing swept angle of −75°. Reynolds number corresponded to one wing chord and is 110,000. Forces and longitudinal moment coefficients are calculated relatively to the total area of all wings, and moment coefficient is calculated relatively to sum of forward wing and rear wing chords. Moment characteristics were defined for conditional center of gravity (CG) that is placed between axes of wings rotation at the distance of 40% from forward wing axis.

3. Results and discussion

Modelling with the help of CFD methods allows getting forces and moments' coefficients, pressure distribution, and pathlines over UAV model for different wings' rotation angles and angles of attack.

Qualitatively, lift coefficient dependences on the angle of attack are analytically predictable as for higher swept angles (lower rotation angles) lift slope is lower, maximal lift coefficient is lower, and critical angle of attack is higher (**Figure 4**).

Calculated UAV polars for different wing rotation angles can be used for relative comparison only (**Figure 5**), as the mesh in boundary layer does not have appropriate density (number of layers and first layer height) to compute drag coefficient properly. So absolute values of drag calculated are not reliable.

Quantitative results prove that UAV is unstable with wings' rotation angles up to 60° (**Figure 6**). Pressure distribution visualization demonstrates that under-pressure zones at upper surfaces of wings are situated ahead of center of gravity (**Figure 7**). So forward and especially rear wings produce lift ahead of center of gravity that results in negative moment and longitudinal static instability. Of course, center of gravity cannot be moved forward because it would be situated ahead of extreme forward CG limit in unfolded position (zero swept).

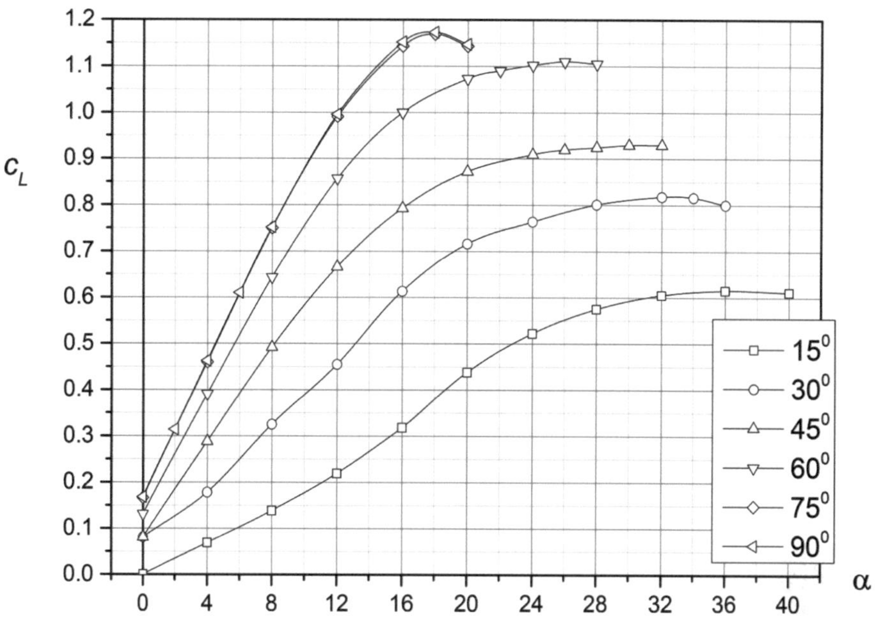

Figure 4. Lift coefficient dependences on angle of attack for different wings' rotation angles.

Obviously, longitudinal instability in wings' rotation angles up to 60° happens not for the given geometry only, but for the different folded tandem-scheme parameters (wingspans ratio, stagger, and height). From the construction point of view, it is rational to fold rear wing ahead and forward wing aft; it is complicated to change design of tube-launches UAV greatly. Telescopic wings may be the solution, but they are associated with design complexity, lower reliability, and airfoil thickness increasing (though bigger profile drag is compensated with high wing aspect ratio and significant induced drag reduction). It is also interesting to rotate forward wings and rear wings differentially (e.g., when forward wings have swept angles of 45°, rear wings have rotation angle of 15°), but obviously rear wing negative pitch moment at rotation angles of 15–30° is higher than positive pitch moment of forward wing (that is also maximal at rotational angles of 15–30°). It may be considered in future research, though the maximal effect of differential wing rotation is the decreasing of instability rotation angles range.

So the most simple practical way is to unfold wings very fast, so the pitch moment does not change the angles of attack significantly.

For high angles of wings' rotation (low sweep angles), UAV model is stable in a wide range of angles of attack because separation occurs mainly from forward wing and from root sections, so the model has no tendencies to roll or to spin. Protection from spin is the advantage of tandem-scheme compared with traditional aircraft, for which separation beginning on root sections does not guarantee spin security. Indeed, if at high angles of attack, pilot or UAV operator deflects the rudder, then sideslip occurs, separation may be eliminated on one wing (e.g., left) and enlarge on the other (right wing), so the aircraft may get into a spin. Nevertheless, for tandem-scheme separation, zone on forward wings would be still much bigger than on rear wings, so the aircraft's nose goes down and away from the dangerous regime.

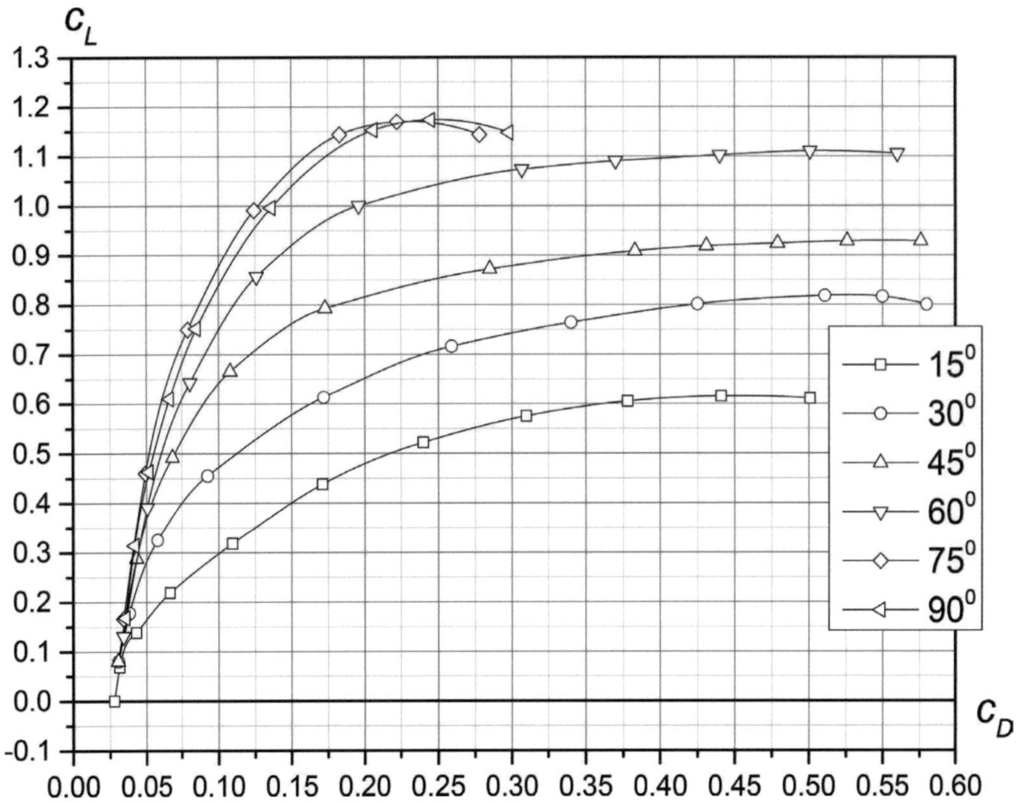

Figure 5. Polars for different wings' rotation angles.

At very high angles of attack, UAV stability is close to neutral (separation occurs on both forward and both rear wings), but the effectiveness of the rear wing ailerons might be reduced due to forward wing tip vortex that increases the angle of attack at tip sections of rear wing that results in small separation areas (**Figure 8**). It does not result in controllability loss, and ailerons on the rear wings are more effective than on forward wings due to larger arm of force. To prevent aileron efficiency, reducing dihedral angles of both wings might be used that decrease negative aerodynamic interference and forward wing tip vortex effect on flow over rear wing (it also can improve lift-drag ratio). The other way is a negative twist of rear wing to decrease local angles of attack for aileron sections.

Local aerodynamic defects were found in the area of the rotation units of both wings (**Figure 9**). Such defects are inherent to folded UAV and may be eliminated with help of flexible fairings that are pressed to the fuselage and wings and slide on their surfaces. However, such installations worsen UAV exploitability because of its permanent deformation and scratches on surfaces and lift-drag ratio remains lower than for UAV without rotation units. It is natural that such fairings were not found on any analogue.

As the folded left and right wings have common axis, and the right wing is situated higher than the left one, there is asymmetry in flow over wings, and a vortex along fuselage's left side is

http://dx.doi.org/10.5772/intechopen.70228

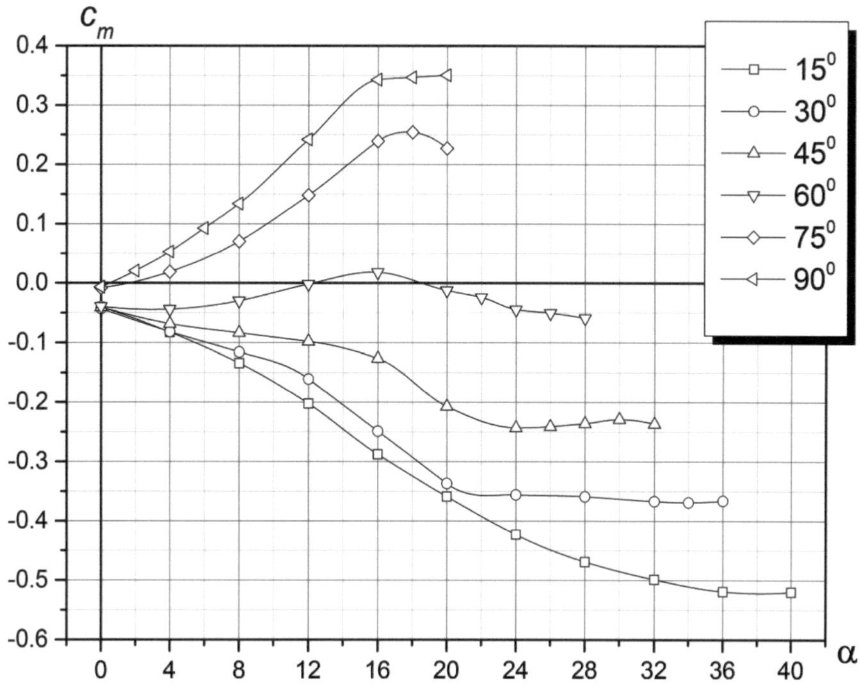

Figure 6. Longitudinal moment coefficient dependences on angle of attack for different wings' rotation angles.

Figure 7. Pressure coefficient distribution (rotation angles 30°, angle of attack 16°).

generated (**Figure 10**). As this vortex is close to symmetry plane, it does not result in a significant roll moment in a wide range of angles of attack (**Figure 11**), and there is no preconditions for UAV spin. The only drawback of this vortex is aerodynamic drag increasing due to flow energy loss.

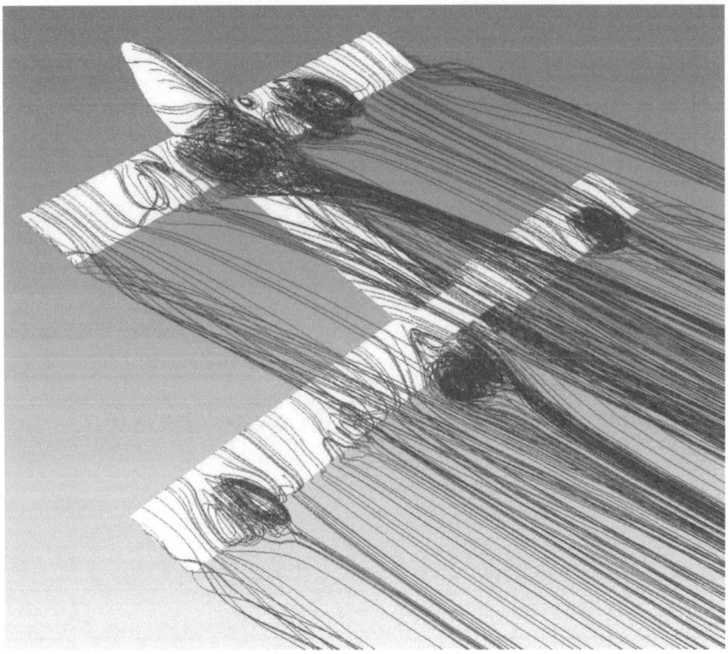

Figure 8. Flow separation from both wings including rear wing aileron sections at high angles of attack.

Figure 9. Flow separation after rotation units at low angles of attack.

Also for low wing rotation angles at middle angle of attack, the flow over rear wings is symmetrical, but the flow differs significantly over forward left and right wings (**Figure 12**). Vortex from wings' rotation unit is located over right wing.

It might be interesting that for low rotation angle wings, the pair of vortices is generated: first one is along leading edge and second one is along trailing edge (**Figure 13**). Rear wings vortices go on forward wings, so one complex vortex system is generated by the whole model.

Figure 10. Vortex along fuselage's left side.

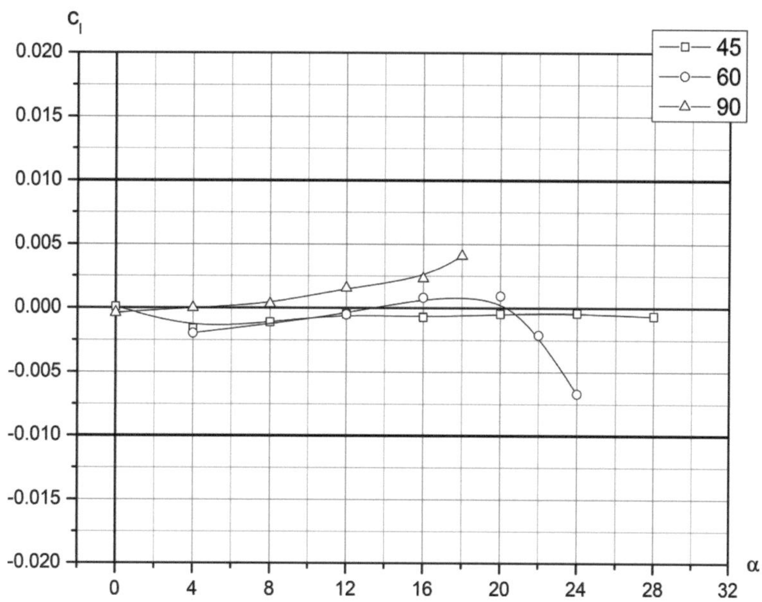

Figure 11. Roll moment coefficient dependences on the angle of attack.

For high angles of attack, separation on rear wings begins at right side because of negative wing-fuselage aerodynamic interference (**Figure 14**). Due to the fact that rear left wing is located lower than the right one, its interference with fuselage is weaker, and there is no

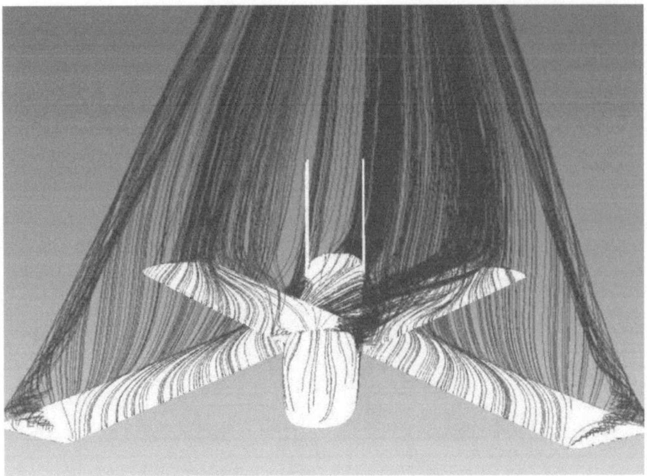

Figure 12. Vortex over forward left wing (rotation angle 15°, angle of attack 24°).

such separation zone. Forward wings are high-wings, so there is no negative interference with the fuselage.

As for tandem-scheme, distance between center of mass and vertical tail is relatively lower than for traditional aerodynamic scheme, so two fins are often necessary to provide yaw stability. As fins also rotate after launch, there are small separation areas after them (**Figure 15**). These aerodynamic defects also may be healed with some flexible fairings as for wing rotation units.

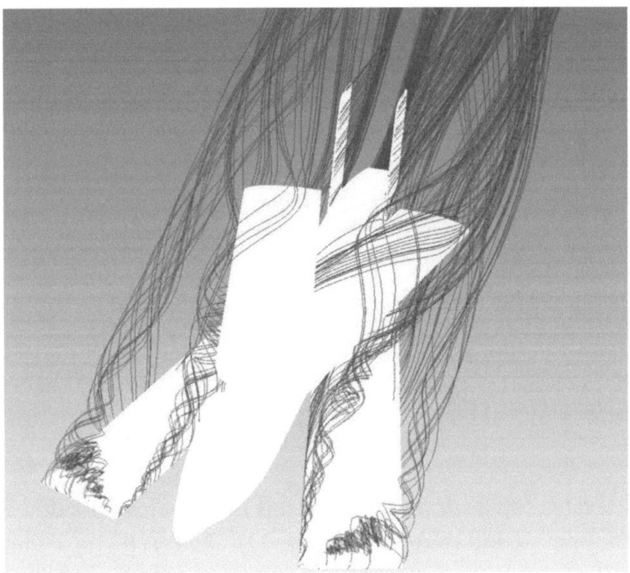

Figure 13. Vortices from rear wings go on forward wings (rotation angle 15°, angle of attack 24°).

Figure 14. Flow separation on root sections of rear right wing (rotation angle 30°, angle of attack 20°).

Figure 15. Flow separation after fin (rotation angle 30°, angle of attack 0°).

4. Conclusion

Quantitative results prove that UAV is unstable with wings' rotation angles up to 60° because rear wings produce lift ahead of center of gravity.

For high angles of wings, rotation (low sweep angles) UAV model is stable in a wide range of angles of attack because separation occurs mainly from forward wing and from root sections, so despite UAV asymmetry, it does not have tendencies to roll or spin.

Local aerodynamic defects were found in the area of the rotation units of both wings. A longitudinal vortex along the left side of fuselage was observed. It is generated due to model asymmetry so that the right wings are higher than the left ones. It was shown that this vortex does not result in significant roll moment.

Due to UAV longitudinal instability for small wings rotation angles, low time of wings unfolding is required.

Further research might include UAV dynamics modelling based on calculated aerodynamics characteristics or flight tests.

Author details

Illia S. Kryvokhatko[1] and Oleksandr M. Masko[2*]

*Address all correspondence to: maska310@ukr.net

1 Antonov Company, Kyiv, Ukraine

2 National Technical University of Ukraine "Igor Sikorsky Kyiv Polytechnic Institute", Kyiv, Ukraine

References

[1] Switchblade: Not a UAV to Mess With [Internet]. March, 10, 2009. Available from: www.deepbluehorizon.blogspot.com/2009/03/switchblade-not-uav-to-mess-with.html [Accessed: 2015-03-09]

[2] BPLA SOKIL-2 [Internet]. Available from: http://shvachko.net/?p=1567&lang=ru [Accessed: 2017-01-28]

[3] Unmanned Integrated Systems. Trident [Internet]. Available from: http://uis.sg/category/uncategorized [Accessed: 2017-01-28]

[4] Zbrutsky OV, Masko OM, Suhov VV. Tube launch unmanned aerial vehicles: Current state and direction research. Journal of Mechanical Engineering the National Technical University of Ukraine «Kyiv Polytechnic Institute». 2012;**64**:63-66. DOI: 10.20535/2305-9001.2012.64.53250

[5] Лапин ЮВ. Статистическая теория турбулентности (прошлое и настоящее - краткий очерк идей) [Statistical theory of turbulence (past and present—brief outline of ideas)]. Научно-технические ведомости. 2004;**2**:7-20

[6] Вождаев ВВ. Влияние модели турбулентности на точность расчета аэродинами ческих характеристик механизированного крыла. [Turbulence model effect on the calculation tolerance of aerodynamic characteristics of wing with high-lift devices]. Техника воздушного флота. 2011;**3**:16-22

Chapter 5

Airfoil Boundary Layer Optimization Toward Aerodynamic Efficiency of Wind Turbines

Youjin Kim, Ali Al-Abadi and Antonio Delgado

Additional information is available at the end of the chapter

http://dx.doi.org/10.5772/intechopen.70895

Abstract

This chapter describes the method of airfoil optimization considering boundary layer for aerodynamic efficiency increment. The advantages of laminar boundary layer expansion in airfoil of horizontal axis wind turbine (HAWT) blades are presented as well. The genetic algorithm (GA) optimization interfaced with the flow solver XFOIL was used with multi-objective function. The power performance of turbine with optimized airfoil was calculated by using blade element method (BEM) in software QBlade. The CFD simulation from OpenFOAM® with Spalart-Allmaras turbulence model showed the visualized airflow. The optimized airfoil shows enlarged laminar boundary layer region in all flow regime with a higher aerodynamic efficiency and the increased gliding ratio (GR). The power velocity and annual energy production (AEP) curves show the performance improvement of wind turbine with the optimized airfoil. The boundary layer thickness and skin-friction coefficient values support the decreased drag of the optimized airfoil. The smaller laminar separation bubbles and reduced stall regime of CFD simulations illustrate the desirable aerodynamics of the resulted airfoil.

Keywords: aerodynamics, airfoil, drag and lift, separation, transition

1. Introduction

The airfoil is the geometrically shaped structure for mechanical force generation from the relative movement between the airfoil and surrounding airflow of the airfoil structures [1]. For wind turbines, the airfoil shape of the blades influences the turbine power production. The lifting efficiency of the blades determines the effectiveness of rotor rotation to cause productive energy conversion from wind kinetics to rotor rotation, which leads to higher electricity generation from the drive unit.

From the late eighteenth century, the curving surface geometry was discovered to be advantageous for lifting efficiency in windmill by Smeaton [2]. In the 1880s, Lilienthal discovered the specific shape from the bird's wings, which inspired the airplane invention by Wright brothers [3]. The research of Prandtl and Tietjens revealed the benefit of the thick airfoil through their mathematical skills and wind tunnel tests in 1917 [4]. The US National Advisory Committee for Aeronautics (NACA)-generated airfoil groups, "NACA airfoil families" in the 1930s, and the series have been widely used even in these days [5]. In contradiction to the mathematical methods to calculate the pressure distribution of airfoil, Jacobs proposed the airfoil design which causes the desired pressure distribution. The laminar flow of airfoil was expanded to cause higher L/D ratio and smaller drag [6]. Later, different types of airfoils for various airplane design and off-design requirements were continuously designed.

The first wind turbine blades were also designed by the airfoils from aeronautic applications. However, in the 1980s, the airfoils specially dedicated for wind turbines were begun to be made due to the defects of aeronautic airfoils applied in a wind turbine. The sensitivity roughness effect on the leading edge arose to be the required element for wind turbine airfoil. The airfoil series for stall-regulated, variable-pitch control wind turbine was developed by NREL in 1984, incorporated with SERI and Airfoils [7]. The wind turbine-dedicated airfoils with the thickness from 15 to 40% of the chord were also made by the team of the Delft University of Technology with the design objective of low sensitivity to roughness, Gurney flaps, and trailing-edge wedge consideration [8]. The airfoils from Risø were designed to have high aerodynamic efficiency and slender blade shape [9]. The airfoil design using numerical optimization for tip region of the blades was researched by Grasso [10]. As mentioned in these studies, the higher aerodynamic efficiency, insensitivity to roughness effect, structural stability and smooth post-stall exhibition, etc., are required for wind turbine airfoil design. To accomplish these objectives, boundary layer consideration of the wind turbine airfoil can be advantageous as it was proven from the laminar airfoil by Jacobs [6].

The boundary layer of the airfoil is exerted by additional pressure generated by the curvature shape of airfoil compared to the constant pressure on boundary layer made of the plate with zero incidences. The pressure distribution on the edge of the boundary layer is same with the pressure distribution on the wall in the plate. However, due to streamline curvature of airfoil surface, the pressure gradients and compensation for the centrifugal force of the streamline flow are generated inside the boundary layer. Furthermore, the transition point of the boundary layer on the airfoil is determined by the outer flow and its pressure difference generated by the curvature shape of the surface [11].

To generate the airfoil shape which has the advantage for pressure distribution in the boundary layer and transition points, genetic algorithm (GA) optimization was used in this study. As all airfoils are designed for higher aerodynamic performance, GA objective functions therefore had two objectives—higher transition points of the larger laminar boundary layer and higher gliding ratio (GR). The airfoil S809 of NREL airfoil series for the wind turbine was chosen as a reference. The shape of insensitiveness to the roughness effect of the airfoil S809 could be maintained in the optimized airfoil. The final evaluations of turbine performance were done with the sample of stall-regulated wind turbine of NREL phase VI, which consists of the same airfoil-type composition [12].

The B-spline parameterization was used for the airfoil description, and the y points of the spline were considered to be the variables. The values of boundary layer parameters and GR of the airfoil were calculated by the flow solver XFOIL. The power performances of turbine unit with blades of the optimized airfoil were calculated by using blade element method (BEM) of the software QBlade. The CFD simulations from OpenFOAM® were performed to visualize the improved aerodynamic aspects.

Section 2 explains the GA airfoil optimization method, Section 3 presents the aerodynamic and boundary layer results of the optimized airfoils with improved power production of turbine unit, Section 4 visualizes the airflow of the optimized airfoil with the reference, and Section 5 concludes this chapter.

2. Genetic Algorithm optimization for airfoil

The GA algorithm is based on the principle of the survival of the fittest and natural selection, observed by Darwin [13]. As the various bird breaks were developed for different foods that they can survive with, the airfoil was set to be shaped to survive at the condition of the highest GR and transition point. To put the airfoil in mathematical form, the B-spline was used, and its variables were set as the y point of control points in MATLAB® (**Figure 1**).

The smoothness and number of variables were set according to the previous research for effective GA operation within a given computation time [14]. The airfoil is described with the B-spline. The corresponding equations and the example of description figure are the following. The x, y points are for the B-spline control points to compose P matrix in *coefs* for determining the smoothness of B-spline with k. When the constant a is designated, the *knots* are defined. The x points of B-spline are defined with MATLAB function *linspace* which divides the x-axis space according to the defined airfoil chord:

$$P = [x; y] \tag{1}$$

$$coefs = [PP] \tag{2}$$

$$knots = -2*a : 2*a + n \tag{3}$$

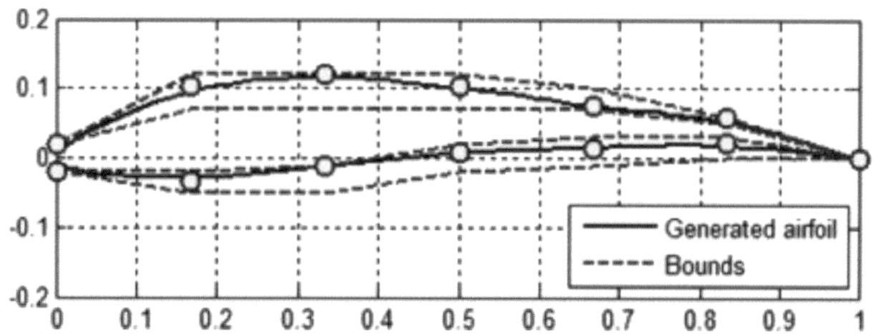

Figure 1. Airfoil B-spline parameterization with upper and lower bounds.

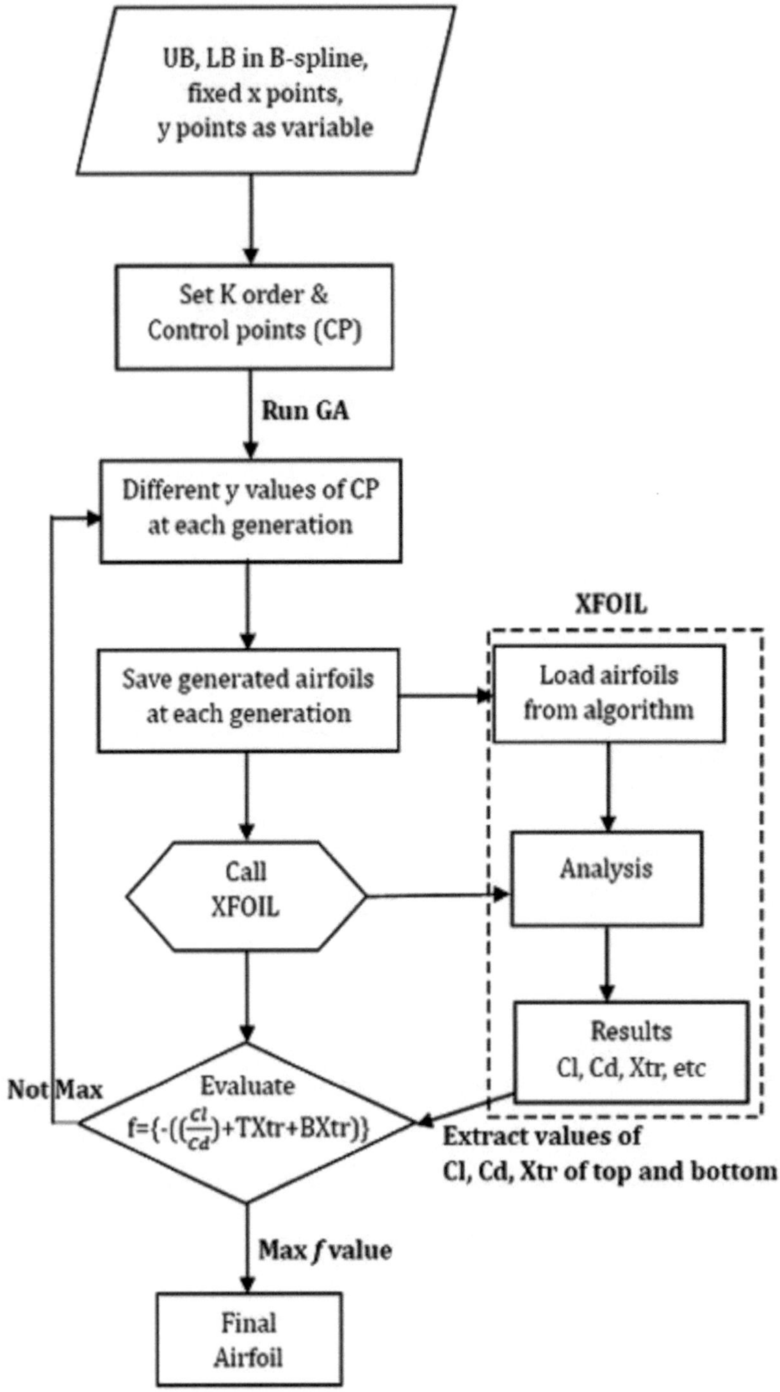

Figure 2. Genetic algorithm flow diagram.

$$k = length(knots) - length(coefs) \quad (4)$$

$$Bsp = spmak(knots, coef) \quad (5)$$

$$x = \left[linspace\left(1, 0, \frac{n}{2}\right), linspace\left(0, 1, \frac{n}{2}\right)\right] \quad (6)$$

To achieve higher aerodynamic efficiency and larger laminar boundary layer region at the same time, the multi-objective function was formulated to evaluate both parameters of the airfoil. The objective function code was written as

$$f = (-(TXtr + BXtr + GR)) \quad (7)$$

Then, *TXtr* is the transition point of the top of airfoil, and *BXtr* is the transition point of the bottom part of airfoil surface. The GR values, *TXtr* and *BXtr* values, were given from the results of XFOIL calculation [15]. The given Re number was 10^6, and the angle of attack was 7°. The algorithm flow diagram is shown in **Figure 2**.

3. The optimized airfoil

The airfoil shape with higher GR and transition points were given after the convergence of default set of population and running operation in GA. The thickness was almost the same as the reference, which was desirable thickness for the stall-regulated wind turbine, but the curvature shape was slightly changed (**Figure 3** and **Table 1**).

However, this slight curvature shape change generated different GR and boundary layer transitions (**Figures 4–8**). As the gliding ratio was set to be higher at the angle of attack 7°, there was some region (off-design) where GR values of the optimized airfoil are lower than the reference. But the angle of attack range of increasing GR ratio was larger at the optimized airfoil (design point). This means that the optimized one has the smaller range of separation or stall occurrence, which is desirable for the stall-regulated wind turbine [7] (**Figure 4**).

Although the maximum GR values of the reference and optimized airfoils were similar, the advantage of the optimized is represented by the drag distribution in **Figure 5**. Significantly, reduced drag coefficients are found in **Figure 5**, which advocate the lower drag of the optimized curvature shape, especially at the stall regime. The reduced drags can be explained with the enlarged laminar boundary layer region of the optimized (**Figures 6–8**).

The transition point distributions of the reference and optimized airfoils are compared in all angle of attack regime. The top and bottom mean the upper and lower sides of the airfoil. Including the angle of attack 7°, where the optimization was calculated, the transition points of the optimized S809gx showed the higher values than the reference S809, which means a larger

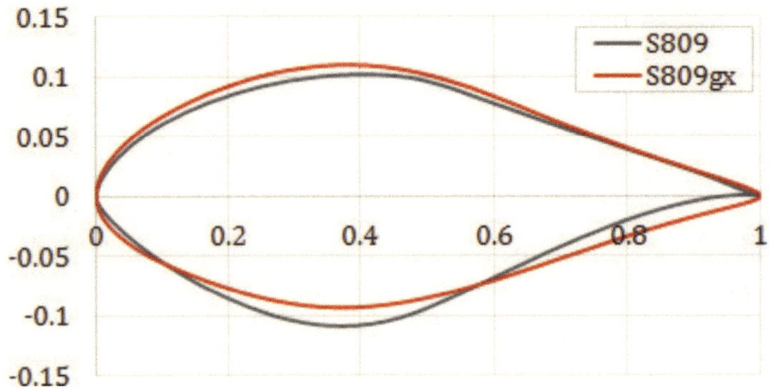

Figure 3. The optimized airfoil S809gx and the reference S809.

	S809	S809gx
Thickness (%)	20.99	20.3
Max. thick. pos. (%)	38.3	38.7
Max. camber (%)	0.99	0.87
Max. camber pos. (%)	83.3	43.6

Table 1. The optimized airfoil S809gx and the reference S809.

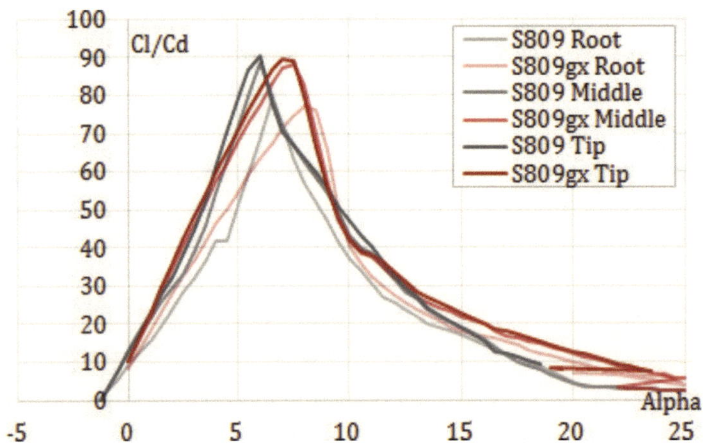

Figure 4. Gliding ratio (Cl/Cd) distribution of the airfoil at the root, middle, and tip parts of the blades.

laminar boundary layer. This tendency was found in all flow regime—fully attached, separation-transition, and dynamic stall regime. The boundary layer region difference can be also found by the skin-friction coefficient (Cf) (**Figure 9**). In the separation-transition flow regime where the tip-speed ratio (TSR) = 5 and the incoming velocity is 8.3 m/s, transition point comparison of two airfoils at the top and bottom parts is presented in **Table 2**. The Cf value at the leading edge of the top of the airfoil showed much higher value than the reference one. The Cf value hill range was also higher at the reference than the optimized one which indicates that the optimized one has the smaller shear wall stress on the top.

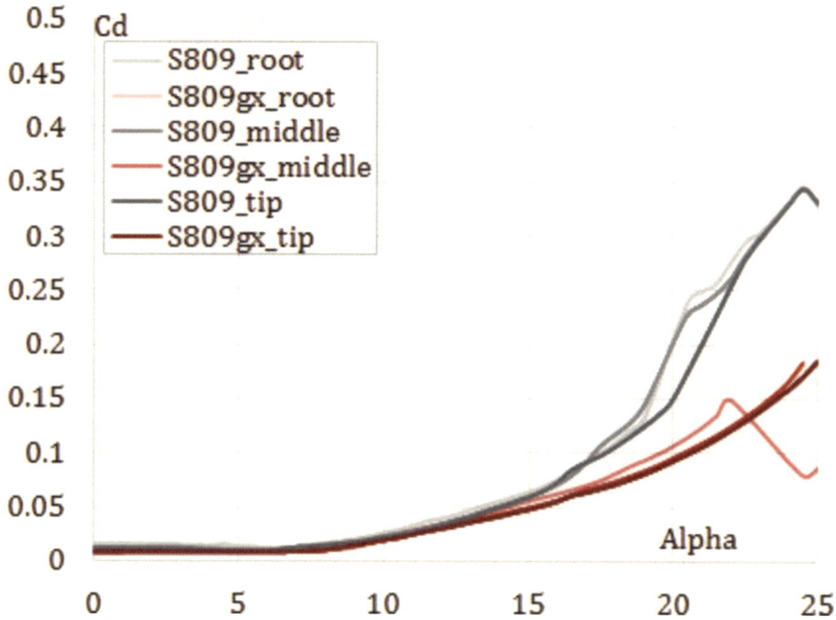

Figure 5. Drag coefficient distribution of airfoils.

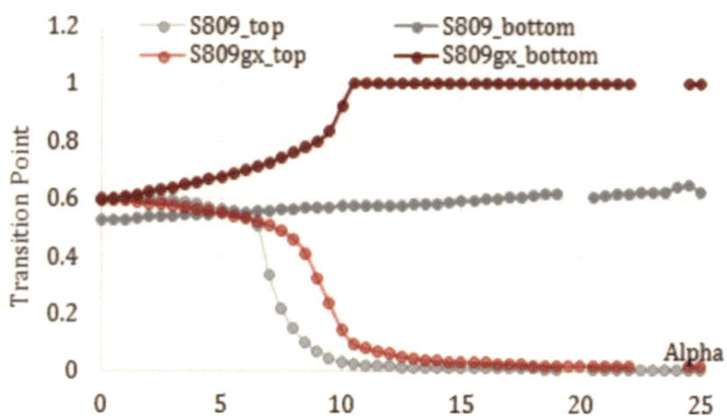

Figure 6. Transition point (Xtr) of the airfoils in fully attached flow regime.

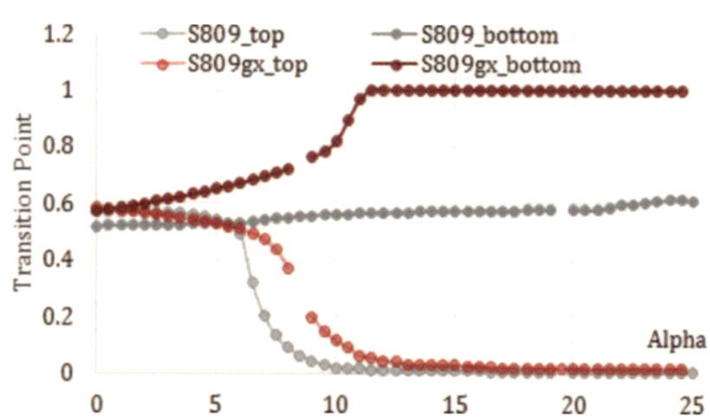

Figure 7. Transition point (Xtr) of the airfoils in transition-separation flow regime.

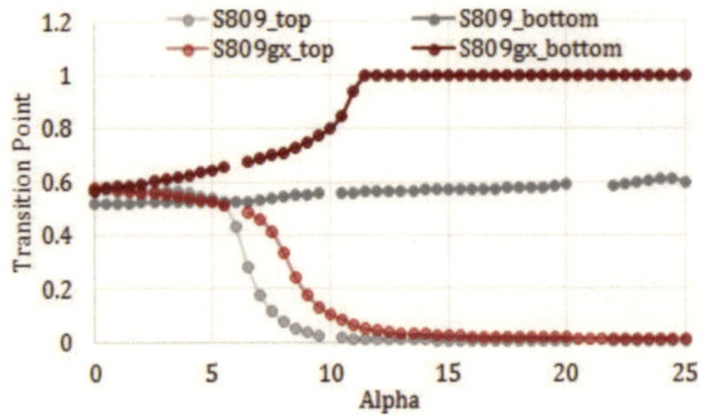

Figure 8. Transition point (Xtr) of the airfoils in dynamic stall flow regime.

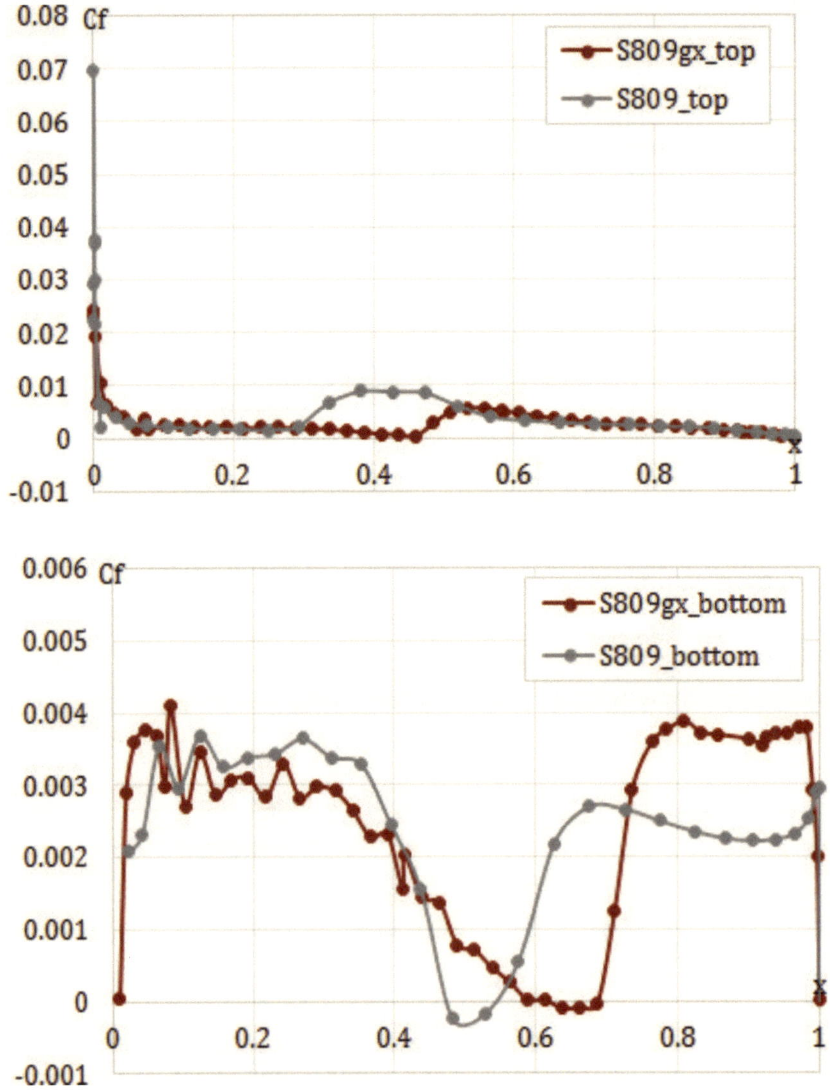

Figure 9. Skin-friction coefficient (Cf) of the top and bottom surfaces of the airfoils.

	S809gx	S809
Xtr_t	0.4589	0.1776
Xtr_b	0.6864	0.5298

Table 2. The optimized airfoil S809gx and the reference S809.

	NREL phase VI with airfoil S809	NREL phase VI with airfoil S809gx
Power regulation	Stall	Stall
Transmission	Single	Single
V cutin/cutout [m/s]	6/25	6/25
Rotational speed [rpm]	71.63	71.63
Outer radius [mm]	5532	5532
Variable losses	0.22	0.22
Fixed pitch/fixed loss	0	0
Weibull setting	k 2(±3) A 9(±3)	k 2(±3) A 9(±3)
Annual yield [W]	49,461,730	59,404,491

Table 3. Settings for the simulation of turbines consist of the airfoil S809 and S809gx.

Although the values were relatively smaller than the top, the bottom part Cf distribution shows the increasing and decreasing tendency change at the transition point. The optimized airfoil shows the changing point to be located to the right side than the reference. If the separation occurs at the point where shear stress gets zero, the optimized airfoil could have the smaller adverse pressure gradient region than the reference one at the bottom side.

The improved aerodynamic efficiency of the optimized airfoil through laminar boundary layer enlargement affected the power increment of rotor turbine with the optimized airfoil. The wind turbine simulation of NREL phase VI blades with the airfoil S809 and S809gx is done with the blade incorporation. The settings for power performance comparison of two turbines are represented in **Table 3** (**Figure 10**). The wind turbine with optimized airfoil shows higher annual yield and power production at a given velocity condition (**Figure 11**). The increased power production means the effective rotor rotation of the wind turbine with the optimized airfoil, which was affected by lifting efficiency of the blades from the aerodynamically upgraded airfoil composition. The 3D rotation effect to reduce the root vortex could be another reason of increased rotor efficiency of the rotor as explained in Ref. [16].

4. CFD simulation

To visualize the improved aerodynamic behavior in the optimized airfoil compared to the reference, the CFD simulations were performed.

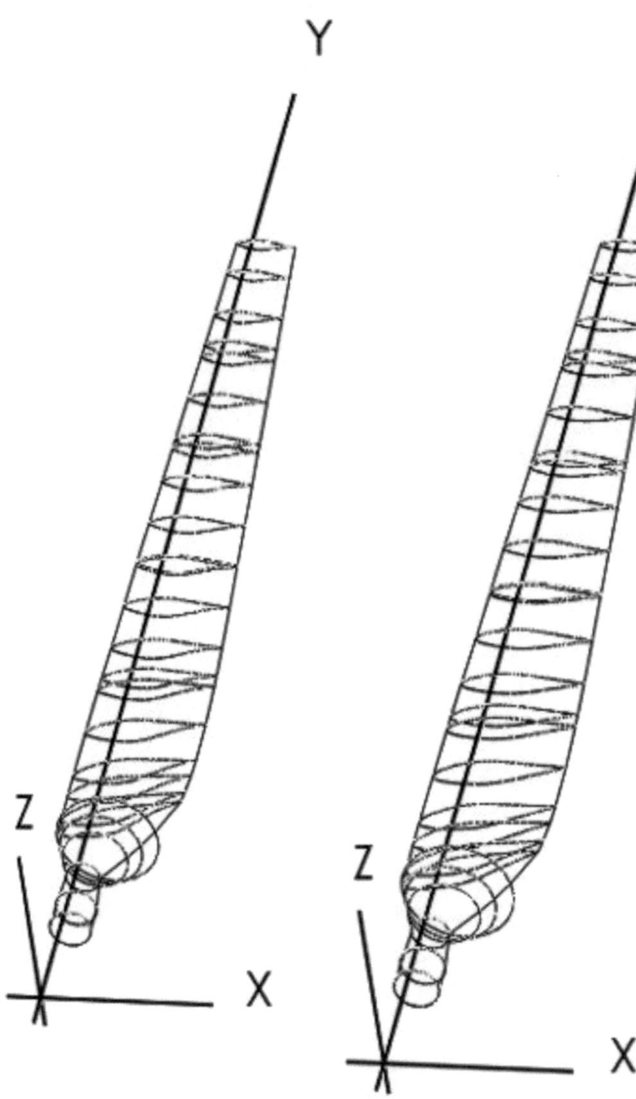

Figure 10. The blade designs with airfoil S809 (left) and S809gx (right).

The meshes for two airfoils were generated with Gmsh [17] (**Figure 12**). The Gmsh tool offers a certain tool for mesh refinement so that the validity of the mesh results can be assured. As the six decimals of accuracy are used, the changes are only appreciated by the fourth digit with the maximum of 0.1% [18].

The software OpenFOAM® uses SimpleFOAM solver with Spalart-Allmaras turbulence model for Reynolds-averaged Navier–Stokes (RANS) equations as the governing equation. It was developed for the aerospace flow problems including wall-bounded flow for boundary layers under the adverse pressure gradients.

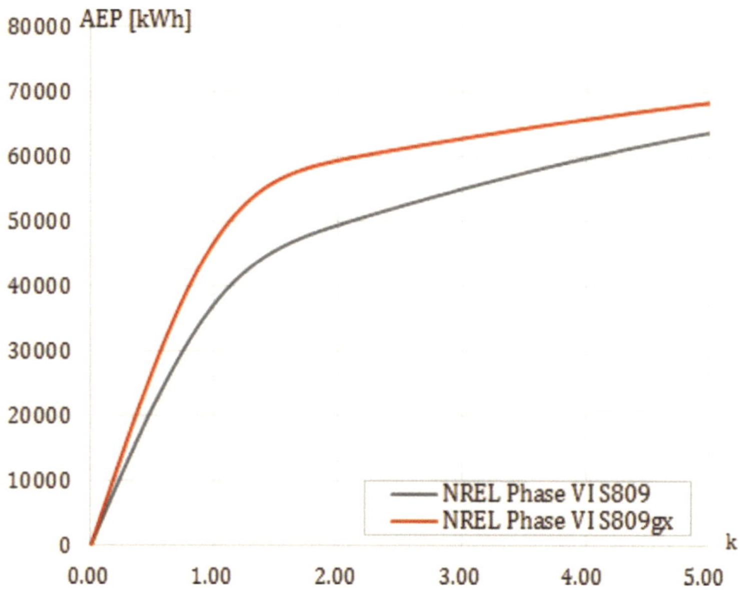

Figure 11. Annual energy production [kWh] of NREL phase VI turbine with the airfoil S809 and S809gx.

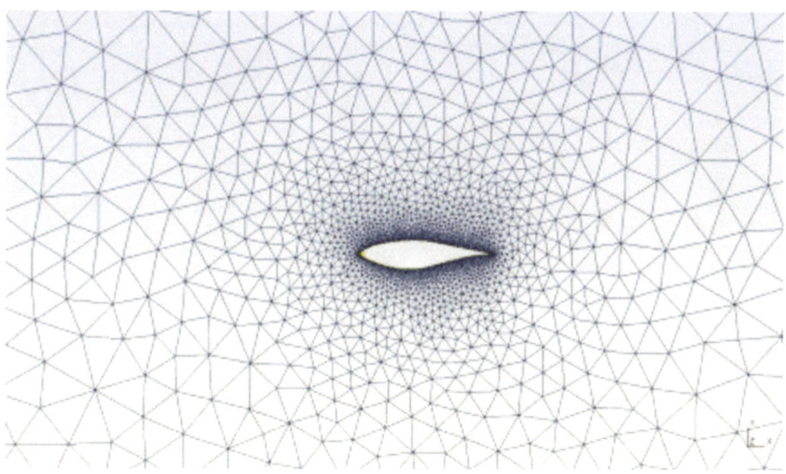

Figure 12. Generated mesh of the reference airfoil S809.

The transport equation with the working variable $\tilde{v}$ is mentioned in Ref. [19]. In model assumption, the eddy viscosity is considered as the capability of turbulent flow to transport momentum. The production term is assumed to be increased linearly with the magnitude of the vorticity (S). The right-hand side of the equation also includes the third term, which is the destruction term. The faster-decaying motion in the outer part of the boundary layer is expressed with the function (f_w). The detailed derivation and explanation about equations are found in Ref. [19].

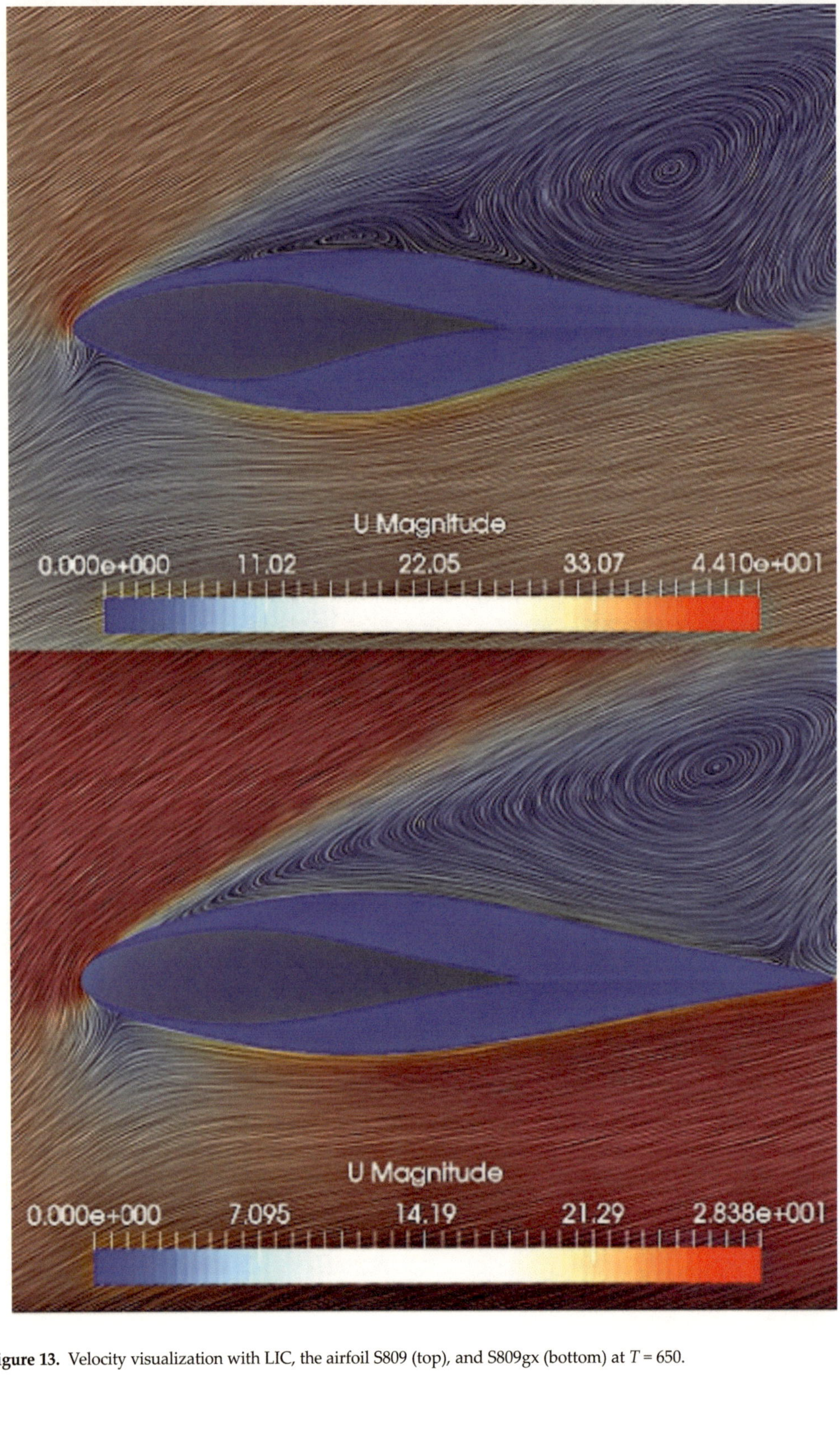

Figure 13. Velocity visualization with LIC, the airfoil S809 (top), and S809gx (bottom) at $T = 650$.

http://dx.doi.org/10.5772/intechopen.70895

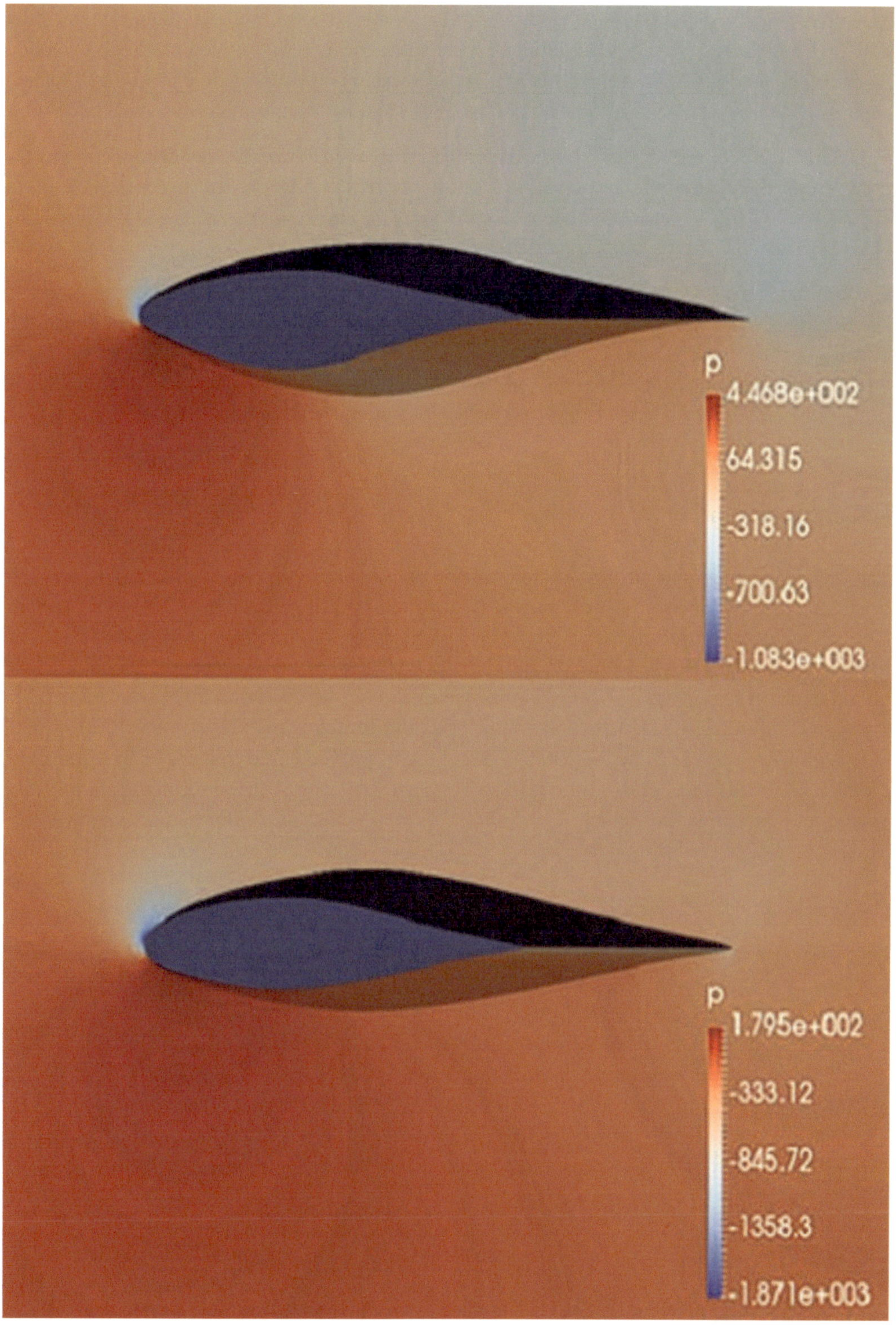

Figure 14. Pressure visualization, the airfoil S809 (top), and S809gx (bottom) at $T = 650$.

The solver visualized the following airflow with ParaView®. The velocity distribution at the outer flow and boundary layer flow distributions are visualized through the aforementioned solution method. The flow time was set to be 1000 with the time interval of 50. The dynamic stall regime angle of attack was at around 22°; the outer flow TSR was 1.5 when incoming velocity was 27.7 m/s.

In general, the reference airfoil shows larger stall area with more laminar separation bubble occurrences than the optimized one. At the time point T = 650, the difference between two airfoils was clearly visualized. At T = 650, the slope of the stall area is more smooth at the optimized airfoil with the smaller total area (**Figure 13**).

The laminar separation bubbles are only found in the reference when the pressure gradient is less drastic at the optimized one. This is supported by the pressure visualization in **Figure 14**. The adverse pressure gradient difference was higher at the reference, and the area with the minus pressure range is largely found at the reference. This milder stall and separation effects explain the smaller drag at the optimized airfoil in dynamic stall regime.

5. Conclusion

The airfoil optimization using GA for higher aerodynamic efficiency and larger laminar boundary layer is achieved. Furthermore, the turbine power performance increment and air flow visualization of reduced stall from the optimized airfoil are proven via simulation. The higher GR and transition point values due to the reduced drag from expanded laminar boundary layer region are presented for the optimized airfoil. The boundary layer characteristics such as diminished skin-friction coefficients and higher transition point values over all flow regimes are found to be the reason behind the improved aerodynamics of the optimized airfoil. The results present the contribution of specifically airfoil shape for aerodynamic efficiency and modified boundary layer distribution on HAWT performance improvement including an effective lifting of the blades and rotor rotation.

Acknowledgements

This study was supported by BB21 project of Busan Metropolitan City (BMC).

Author details

Youjin Kim[1,2]*, Ali Al-Abadi[3] and Antonio Delgado[1,2]

*Address all correspondence to: youjin.kim@fau.de

1 Institute of Fluid Mechanics, FAU Erlangen-Nürnberg, Germany

2 Institute of Fluid Mechanics, FAU Erlangen-Nürnberg Busan Campus, South Korea

3 SGB Power Transformer, Regensburg, Germany

References

[1] James F. Manwell, Jon G. McGowan, Anthony L. Rogers. Wind Energy Explained: Theory, Design and Application,. 2nd ed. UK: Wiley; 2009. 704 p. ISBN: 978-0-470-01500-1

[2] Smeaton J. Reports of the Late John Smeaton, F.R.S., Made on Various Occasions, in the Course of his Employment as a Civil Engineer. 2nd ed. M. Taylor: London, UK; 1837 556 p

[3] Otto Lilienthal. Birdflight as the Basis of Aviation: A Contribution Towards a System of Aviation. 1st ed. United States: Markowski International Publishers; 2000. 151 p. ISBN: 0938716581

[4] Prandtl L, Tietjens OG. Applied Hydro- and Aeromechanics (Dover Books on Aeronautical Engineering). United States: Dover Publications; 2012. 336 p 048660375X

[5] Michael J. Rycroft, Wei Shyy Editors. A History of Aerodynamics: And its Impact on Flying Machines. Cambridge Aerospace Series. new ed. United Kingdom: Cambridge University Press; 2012. 336 p. ISBN: 048660375X

[6] Jacobs, Eastman N. (National Advisory Committee for Aeronautics). NASA Technical Reports Server [Internet]. 1939 [Updated: 1996]. Available from: https://ntrs.nasa.gov/search.jsp?R=19930092782

[7] Tangler JL, Somers DM. NREL airfoil families for HAWT's. In: WIND POWER. Washington, DC; Colorado, U.S: National Renewable Energy Laboratory (NREL); 1995; p. 117–123

[8] Timmer WA, van Rooij RPJOM. Summary of the Delft University wind turbine dedicated airfoils. In: ASME 2003 Wind Energy Symposium; January 6–9, 2003; Reno, Nevada, USA. ASME; 2003. WIND2003–352, pp. 11–21. DOI: 10.1115/WIND2003-352

[9] Fuglsang P, Bak C. Design and verification of the NewRisø-A1 airfoil family for wind turbines. In: A Collection of the 2001 ASME Wind Energy Symposium Technical Papers at the 39. AIAA Aerospace Sciences Meeting and Exhibit. AIAA-2001-0028. United States: American Institute of Aeronautics and Astronautics; American Society of Mechanical Engineers; 2001. p. 81–91. ISBN: 1-56347-476-X

[10] Grasso F. Usage of numerical optimization in wind turbine Airfoil design. In: Proceedings, 28th AIAA Applied Aerodynamics Conference. Journal of Aircraft; 28 June-1 July 29 2010; Chicago, IL, USA; The American Institute of Aeronautics and Astronautics (AIAA); 2011. 2010–4404

[11] Schlichting H, Gersten K. Boundary-Layer Theory. Berlin, Heidelberg: Springer; 2017. 805 p. DOI: 10.1007/978-3-662-52919-5

[12] Muljadi E, Pierce K, Migliore P. Control strategy for variable-speed, stall-regulated wind turbines. In: 1998 American Controls Conference; 06/24/1998-06/26/1998; Philadelphia, PA (US). Golden, Colorado: National Renewable Energy Lab., Golden, CO (United States); 1998. ark:/67531/metadc690735

[13] Mitsuo Gen, Runwei Cheng. Genetic Algorithms and Engineering Optimization. Copyright (C) 2000 John Wiley & Sons, Inc. Published Online: 12 NOV 2007; ISBN: 978-0-471-31531-5; p. 512

[14] Kim Y, Al-Abadi A, Delgado A. Strategic blade shape optimization for aerodynamic performance improvement of wind turbines. In: ASME Turbo Expo 2016: Turbomachinery Technical Conference and Exposition, Volume 9: Oil and Gas Applications; Supercritical CO_2 Power Cycles; Wind Energy; June 13–17, 2016; Seoul, South Korea. ASME; 2016. Paper No. GT2016–56836. pp. V009T46A009. 10 pages. DOI: 10.1115/GT2016-56836

[15] Mueller TJ. Low Reynolds Number Aerodynamics. 1st ed. Berlin Heidelberg: Springer-Verlag; 1989. 451 p. DOI: 10.1007/978-3-642-84010-4

[16] Bangga G, Lutz T, Jost E, Kraemer E. CFD studies on rotational augmentation at the inboard sections of a 10 MW wind turbine rotor. Journal of Renewable and Sustainable Energy. 2017;**9**:023304 http://dx.doi.org/10.1063/1.4978681

[17] Christophe Geuzaine and Jean-François Remacle. A three-dimensional finite element mesh generator with built-in pre- and post-processing facilities [Internet]. [Updated: version 3.0.3, June 23 2017]. Available from: http://gmsh.info

[18] Sánchez SJR. Analysis of flow separation over aerodynamic airfoils [thesis]. Madrid, Spain: Universidad Carlos III De Madrid Escuela Politécnica Superior; 2014. 72 p Available from: https://core.ac.uk/download/pdf/44310532.pdf

[19] Deck S, Duveau P, d'Espiney P, Guillen P. Development and application of Spalart-Allmaras one equation turbulence model to three-dimensional supersonic complex configurations. Aerospace Science and Technology, DOI. 2002;**6**(3):171-183 https://doi.org/10.1016/S1270-9638(02)01148-3

Chapter 6

LES of Unsteady Aerodynamic Forces on a Long-Span Curved Roof

Wei Ding

Additional information is available at the end of the chapter

http://dx.doi.org/10.5772/intechopen.70880

Abstract

The present paper discusses the unsteady aerodynamic forces on long-span curved roofs by using large eddy simulation (LES). The forced vibration test in a turbulent boundary layer is simulated. The models are force vibrated in the first anti-symmetric mode to investigate the influences of a roof's vibration on the wind pressure and flow field around a vibrating roof. The characteristics of unsteady aerodynamic forces in a wider range of reduced frequency of vibration are also investigated. A comparison between the wind tunnel experiment and the LES indicates that the LES can be used effectively to evaluate the unsteady aerodynamic force.

Keywords: large eddy simulation, unsteady aerodynamics force, long-span curved roof, forced vibration test

1. Introduction

Wind-structure interaction is a critical consideration in the design of many structures in civil engineering, especially for structures being flexible and light, such as long-span bridges, high-rise buildings, and long-span roofs. Such structures are generally vulnerable to the dynamic wind actions because of low damping and frequency [1–4]. The wind-structure interaction induces unsteady aerodynamic forces, or motion-induced wind forces, which may affect the wind-induced response significantly [5, 6]. Therefore, the unsteady aerodynamic force is an important consideration in the design of long-span vaulted roofs. Uematsu and Uchiyama [7] conducted a series of wind tunnel tests using elastic models of a one-way type of suspended roof. The mechanism of the wind-induced vibrations and the effect of wind-roof interaction on the dynamic response were discussed. Daw and Davenport [8] carried out a forced vibration test on a semi-circular roof to investigate the dependence of unsteady aerodynamic forces on the turbulence intensity, wind speed, vibration amplitude, and geometric details of the roof.

Ohkuma *et al.* [9] investigated the mechanism of aeroelastic instability of long-span flat roofs using a forced vibration test in a wind tunnel. At present, long-span curved roofs are universally constructed. However, there is an insufficient research on unsteady aerodynamic forces on long-span curved roofs, and the characteristics of unsteady aerodynamic forces are not well understood. Therefore, it is necessary to investigate this problem further for proposing more reasonable methods of response analysis for these roofs.

In this chapter, we investigate the characteristics of unsteady aerodynamic forces acting on long-span curved roofs for improving the wind-resistant design method. The large eddy simulation (LES) is used to discuss the influences of a roof's vibration on the wind pressure and flow field around a vibrating roof. The characteristics of unsteady aerodynamic forces in a wider range of reduced frequency of vibration are also investigated. The results of LES are validated by comparing with the experimental results.

2. Unsteady aerodynamic force

This section is focused on the illumination of unsteady aerodynamic forces (motion-induced forces), which result from the wind-structure interaction. Fluctuating deflections of the structure may be excited by the turbulence in oncoming flow, or the wake instability caused by vortex shedding in the structural wake. The unsteady aerodynamic forces result from the modification of the flow as the structure vibrates or changes shape, in other words, the interaction of the wind flow and structure. These forces may lead to instability. The unsteady aerodynamic force is described as two components: the aerodynamic stiffness term that is in-phase with the displacement and the aerodynamic damping term that is out-phase with displacement.

The aerodynamic stiffness is the added stiffness of the air surrounding the structure, which may increase or effectively reduce available structural static stiffness. For a conventional heavy structure, the aerodynamic stiffness is generally insignificant in comparison to the structural stiffness. However, for a long-span light-weight structure, which vibrates more easily in the wind, the aerodynamic stiffness may change the structural response. For instance, if the total static stiffness of the system in wind is reduced to zero, then a divergent instability may be induced.

When a structure is vibrating in the wind, the relative velocity of the structure to the wind flow changes in magnitude and direction. This phenomenon effectively produces an added damping force, referred to as aerodynamic damping. The aerodynamic damping may add to the structural damping to reduce the response of structure, or become negative and increase the response of the structure. The chances of aerodynamic instability are high as the total damping in the system approaches zero.

2.1. Definition of unsteady aerodynamic force

The displacement of structure in the jth mode may be represented by the following equation,

$$Z_j(s,t) = \varphi_j(s)x_j(t) \tag{1}$$

where φ_j and x_j are the mode shape and generalized displacement of the jth mode, respectively; and s represents the circumferential coordinate taken along the roof.

Applying a modal analysis to the equation of motion for the roof, we obtain the following equation of motion for the *j*th generalized displacement,

$$M_{s_j}\ddot{x}_j(t) + C_{s_j}\dot{x}_j(t) + K_{s_j}x_j(t) = F_{W_j}(t) + F_{A_j}(t) \tag{2}$$

$$\ddot{x}_j(t) + 2\xi_{s_j}\omega_{s_j}\dot{x}_j(t) + \omega_{s_j}^2 x_j(t) = F_j(t)/M_{s_j} \tag{3}$$

$$F_j(t) = F_{W_j}(t) + F_{A_j}(x, \dot{x}, \ddot{x}\ldots) \tag{4}$$

where M_{sj} = generalized mass, ω_{sj} = natural circular frequency, ζ_{sj} = critical damping ratio, and F_j = generalized force. F_{Wj} represents the fluctuating wind force due to the oncoming flow and wake instability, while F_{Aj} represents the unsteady aerodynamic force due to the wind-roof interaction.

In the case of the forced-vibration test, a steady vibration in the first anti-symmetric mode represented by a sine curve is applied to the roof. The unsteady aerodynamic force F_{Aj} (here $j = 1$) can be obtained from Eq. (5) by using the Fourier series at the frequency f_m of the forced vibration:

$$F_{A_j}(t) = F_{R_j}\cos 2\pi f_m t - F_{I_j}\sin 2\pi f_m t \tag{5}$$

$$F_{R_j} = \frac{1}{T}\int_{-T}^{T} F_j(t)\cos 2\pi f_m t \tag{6}$$

$$F_{I_j} = \frac{1}{T}\int_{-T}^{T} F_j(t)\sin 2\pi f_m t \tag{7}$$

where F_{Rj} and F_{Ij} are the in-phase and out-of phase components of the unsteady aerodynamic force, respectively.

3. Large eddy simulation

The LES is used to investigate the characteristics of unsteady aerodynamic forces. The influences of a roof's vibration on the wind pressure and flow field around a vibrating roof are also investigated. The simulation is carried out by using a CFD software 'STAR-CD'.

3.1. Computational outline

3.1.1. Computational model

The computational model used in the 'STAR-CD' is shown in **Figure 1**. In order to investigate the effect of geometric shape on the unsteady aerodynamic force, the rise/span ratio *r*/*L* of computational models is assumed to be 0.15, 0.20, and 0.25. The curved roof model is forced to vibrate in the first anti-symmetric mode as shown in **Figure 1**.

3.1.2. Computational parameters

Table 1 summarizes the computational parameters. In order to discuss the effect of geometric shape on wind-roof interaction, the rise/span ratio is changed from 0.15 to 0.25. The amplitude

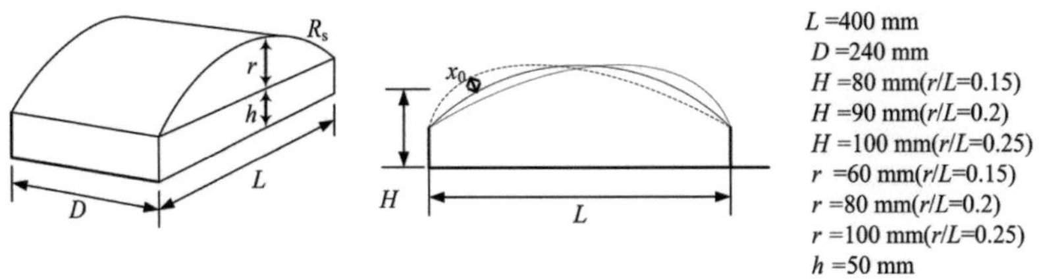

Figure 1. Computational model.

Wind speed	5 m/s
Forced vibration amplitude	4 mm
Rise/span ratio (r/L)	0.15, 0.20, 0.25
Forced vibration frequency (f_m)	0–160 Hz (10 Hz increment)
Reduced vibration frequency(f_m^*)	0–2.5

Table 1. Parameter of CFD simulation.

x_0 of vibration is fixed to 4.0 mm (i.e. $x_0/L = 1/100$). In this study, based on the assumptions of the mean roof height for real structure $H_{\_r}$ = 20 m; the wind speed at mean roof height U_{H_r} = 20–40 m/s; the natural frequency f_s = 0.4–2.5 Hz. We calculated the reduced frequency for real roof $f_{\_r}^*$ = 0.2–2.5, as shown in **Table 3**. In order to satisfy the similarity principle of real long-span roofs $f_m^* = f_{\_r}^*$ (f_m^* = the reduced frequency for model), the forced vibration frequency f_m should be set at 12.5–156.25 Hz, as shown in **Table 2**. With regard to the limitation of forced vibration equipment used in the wind tunnel experiment [10], the forced vibration

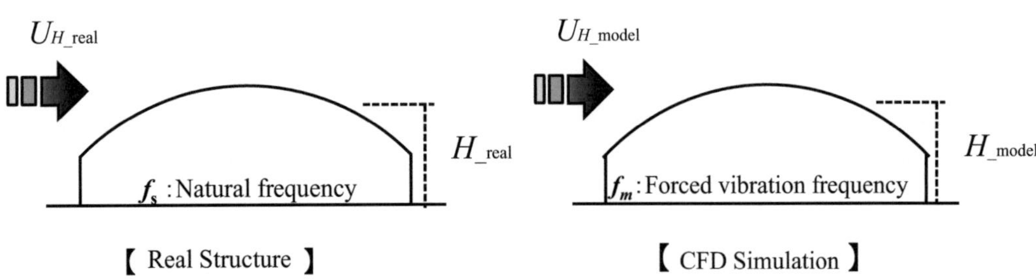

$f_{\_r}^* = \frac{f_s H_{\_r}}{U_{H\_r}}$	$f_m^* = \frac{f_m H}{U_H}$
Mean roof height for real structure $H_{\_r}$ = 20 m	Mean roof height for model H = 0.08 m
Wind speed at mean roof height U_{H_r} = 20–40 m/s	Wind speed at mean roof height U_H = 5 m/s
Natural frequency f_s = 0.4–2.5 Hz	Forced vibration frequency f_m = 12.5–156.25 Hz
Reduced frequency for real roof $f_{\_r}^*$ = 0.2–2.5	Reduced frequency for model $f_m^* = f_{\_r}^*$ = 0.2–2.5

Table 2. Determination of forced vibration frequency.

frequency cannot be set as large as this. Therefore, it is necessary to use LES to examine the characteristics of unsteady aerodynamic forces in a wider range of reduced frequency. For the LES, we change the forced vibration frequency from 0 to 160 Hz and the range of reduced frequency of vibration is from 0 to 2.5, as shown in **Table 1**.

3.1.3. Computational domain

Figure 2 shows the computational domain. In this study, the length of span direction equals the span of roof to generate two-dimensional flow that is corresponded with that used in the wind tunnel experiment.

3.1.4. Computational mesh

In the simulation, various types of mesh arrangements were calculated. We compared the results of LES with those of wind tunnel experiment. And then, the mesh arrangement was selected which leads to the most corresponding results with that of experiment, as shown in **Figure 3**. The magnitude of minimum mesh is 0.15×10^{-3}. And the dynamic mesh is used to simulate the vibration of model.

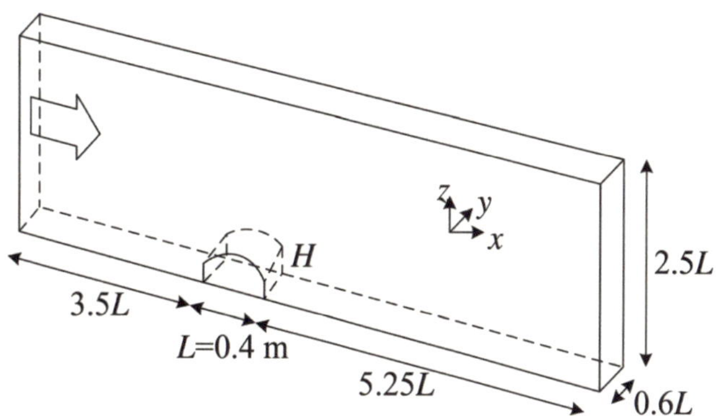

Figure 2. Computational domain.

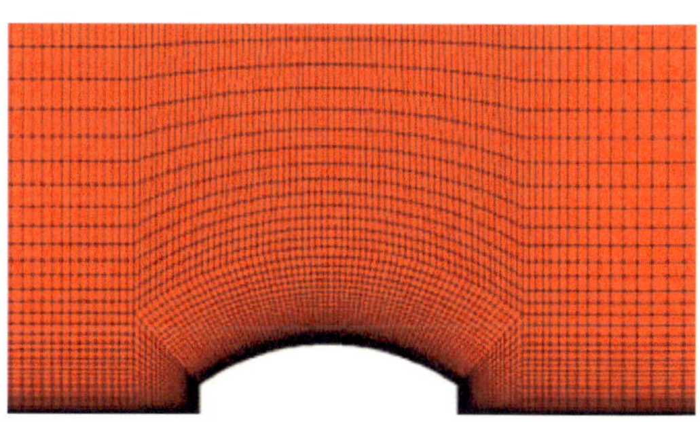

Figure 3. Mesh arrangement around roof.

3.1.5. Computational and boundary conditions

There are mainly three types of CFD approaches, which are used in computational wind engineering (CWE): the Reynolds-averaged Navier-Stokes (RANS), the large eddy simulations (LES), and direct numerical simulation (DNS). Due to the limitation of available computer memory and speed at present, DNS cannot be widely used in CWE for solving complicated practical problems. RANS solves the time-averaged NS equations, and the averaged solution reflects the averaged properties of the turbulent flow. Thus, the time-averaged solution is less trustable in nonstationary flows. On the other hand, LES resolves the scale of motion larger than the gird size and the effect of motion of turbulent eddy smaller than grid scale needs to be modeled. The unsteady motions of large eddy can be explicitly predicted and the accuracy is usually much better than RANS models, since the effects of only small eddy are modeled. Therefore, the LES is adopted in this study.

The governing equations adopted in the present LES method are the spatially filtered continuity and Navier-Stokes equations as follows,

$$\frac{\partial \overline{u}_i}{\partial x_i} = 0 \tag{8}$$

$$\frac{\partial \overline{u}_i}{\partial t} + \frac{\partial \overline{u}_i \overline{u}_j}{\partial x_j} = -\frac{1}{\rho}\frac{\partial \overline{p}}{\partial x_i} + \frac{\partial}{\partial x_j}\left(\nu \frac{\partial \overline{u}_i}{\partial x_j}\right) - \frac{\partial \tau_{ij}}{\partial x_i} \tag{9}$$

$$\tau_{ij} = \overline{u_i u_j} - \overline{u}_i \overline{u}_j \tag{10}$$

where u_i, u_j represent flow velocity in *i*-direction and *j*-direction. The p, ρ, and υ represent pressure, density, and dynamic viscosity of the fluid, respectively. The $(\overline{u})$ denotes application of the spatial filter. The τ_{ij} is subgrid-scale (SGS) stress, which is parameterized by an eddy viscosity model. The standard Smagorinsky model is adopted to estimate the term of τ_{ij}, as shown in Eqs. (11)–(13).

$$\tau_{ij} = -2\nu_{SGS}\overline{S}_{ij} + \frac{1}{3}\delta_{ij}R_{kk} \tag{11}$$

$$\overline{S}_{ij} = \frac{1}{2}\left(\frac{\partial \overline{u}_i}{\partial x_j} + \frac{\partial \overline{u}_j}{\partial x_i}\right) \tag{12}$$

$$\nu_{SGS} = \left(C_s\overline{\Delta}\right)^2\sqrt{2\overline{S}_{ij}\overline{S}_{ij}} \tag{13}$$

where υ_{SGS} is the subgrid-scale turbulent eddy viscosity. $\overline{S}_{ij}$ is the rate of strain tensor for the resolved scale. Δ means the spatial filter width and C_s is the Smagorinsky constant and is taken as 0.12 (C_s = 0.12). The computational and boundary conditions are summarized in **Table 3**.

Computational domain	9.75L(x) × 0.6L(y) × 2.5L(z)
Inlet boundary	Inflow turbulence is generated in preliminary computational domain
Upper boundary	Zero normal velocity and zero normal gradients of other variables
Side boundary	Cyclic boundary conditions
Outlet boundary	Zero normal gradients of all variables
Floor and model surfaces	No-slip condition
Grid discretization	260(x) × 24(y) ×64(z)=(399, 360)
Convection schemes	Second-order centered difference scheme
Time differential schemes	Euler implicit
Numerical algorithm	PISO algorithm
Time	T = 4 s, Δt = 2.0E−04 s (Courant Number: 9.1E−02)

Table 3. Computational and boundary conditions.

3.1.6. *Inflow turbulence*

As is known, the flow around a structure is strongly affected by the flow turbulence. Therefore, the proper generation of the inflow turbulence for the LES is essential in the determination of wind loads on structures. At present, several techniques have been developed. In general, there are three kinds of inflow turbulence generation methods. The first approach is to store the time history of velocity fluctuations obtained from a preliminary LES computation. Nozu and Tamura [11] employed the interpolation method with the periodic boundary condition to simulate a fully developed turbulent boundary layer and tried to change the turbulent characteristics by using roughness blocks. Another approach is to numerically simulate the turbulent flow in auxiliary computational domains (often called a driver region set at the upstream region of a main computational domain). Lund *et al.* [12] proposed the method to generate turbulent inflow data for the LES of a spatially developing boundary layer. Kataoka and Mizuno [13] simplified Lund's method by assuming that the boundary layer thickness is constant within the driver region, and only the fluctuating part of velocity is recycled in the streamwise direction. Nozawa and Tamura [14, 15] discussed the potential of large eddy simulation for predicting turbulence characteristics in a spatially developed turbulent boundary layer over a rough ground surface and improved Lund's method. The third approach is to use artificial numerical models to generate inflow turbulence statistically [16–19].

In this study, we use a preliminary LES to simulate inflow turbulence and store the time history of velocity fluctuations. **Figure 4** shows a schematic illustration of the domain of the preliminary computation. In the domain, the roughness blocks with heights 3, 5, and 8 cm are distributed on the ground to generate turbulence. The computational and boundary conditions are summarized in **Table 4**.

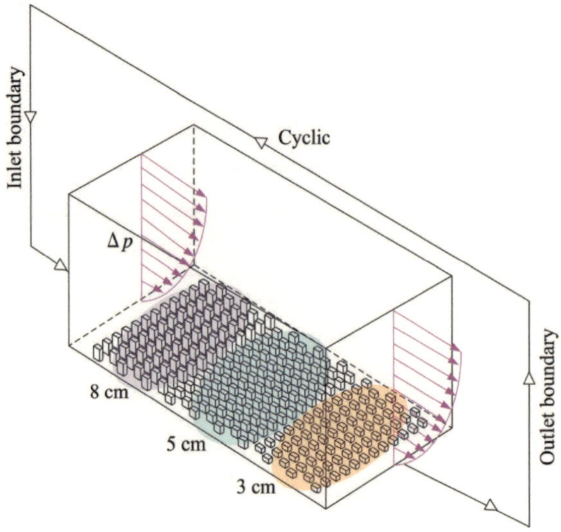

Figure 4. Preliminary computational domain.

Inlet boundary	Cyclic boundary condition
Upper boundary	Zero normal velocity and zero normal gradients of other variables
Side boundary	Cyclic boundary conditions
Outlet boundary	Cyclic boundary conditions
Floor and surfaces of roughness blocks	No-slip condition
Convection schemes	MARS method
Diffusion schemes	Centered difference scheme
Time differential schemes	First order Euler implicit
Numerical algorithm	PISO algorithm
Time step	$\Delta t = 2.0E{-}04$ s

Table 4. Computational and boundary conditions.

The profiles of the mean wind speed and turbulent intensity at the inlet of the computational domain are shown in **Figure 5(a)**. The longitudinal velocity spectrum at a height of $H = 90$ mm is shown in **Figure 5(b)**. In both figures, the results of wind tunnel flow are also plotted for comparative purposes. It can be seen that the inflow turbulence used in the LES is generally in good agreement with that used in the wind tunnel experiment.

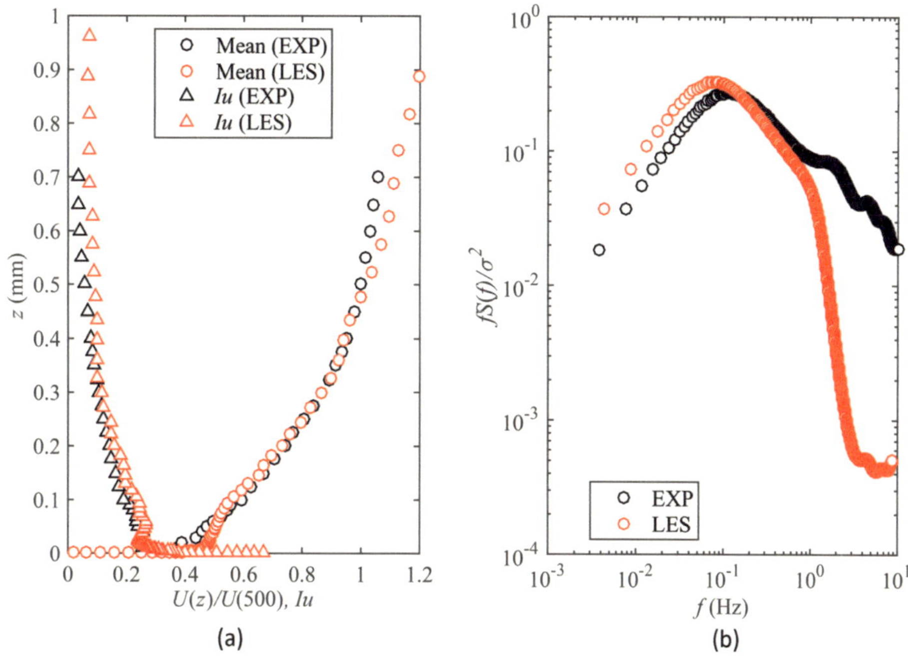

Figure 5. Comparison of inflow turbulence between wind tunnel experiment and LES; (a) profiles of mean wind speed ($U(z)$) and turbulence intensity (Iu); (b) longitudinal velocity spectrum (H = 90 mm).

4. Results and discussion

4.1. Comparison with wind tunnel experiment

In order to validate the LES computation, the distributions of the mean wind pressure coefficient C_{p_mean} and fluctuating wind pressure coefficient C_{p_RMS} along the centerline of the vibrating roof is compared with those obtained from the wind tunnel experiment. **Figures 6** and **7** show the results, in which the results for the frequencies of 0, 10, and 15 Hz are plotted. It can be seen that there is generally a good agreement between the LES and the wind tunnel experiment. In **Figure 6**, the difference is somewhat larger near the rooftop; the LES values are approximately 10% larger in magnitude than the experimental ones. This difference may be due to a difference in surface roughness of the roof between the LES and the wind tunnel experiment. In **Figure 7**, when the f_m = 0 Hz, the value of C_{p_RMS} for the LES is larger than that for the wind tunnel test. That maybe because that the turbulence intensity of inflow turbulence used in the LES is slightly larger than that used in the wind tunnel test (see **Figure 5**).

4.2. Distribution of wind pressure on the roof

The distributions of mean and rms fluctuating wind pressure coefficients for various forced-vibration frequencies are shown in **Figure 8**. It can be seen that the mean wind pressure

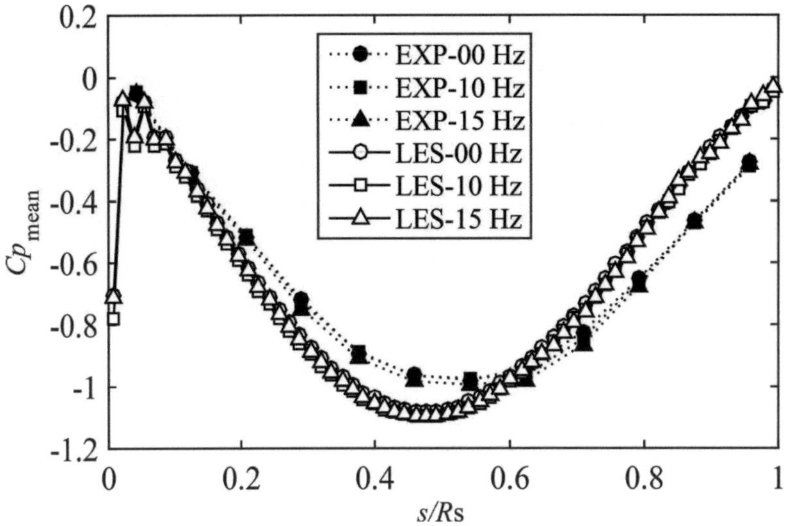

Figure 6. Comparisons for the distribution of the mean wind pressure coefficients along the centerline between LES and wind tunnel experiment.

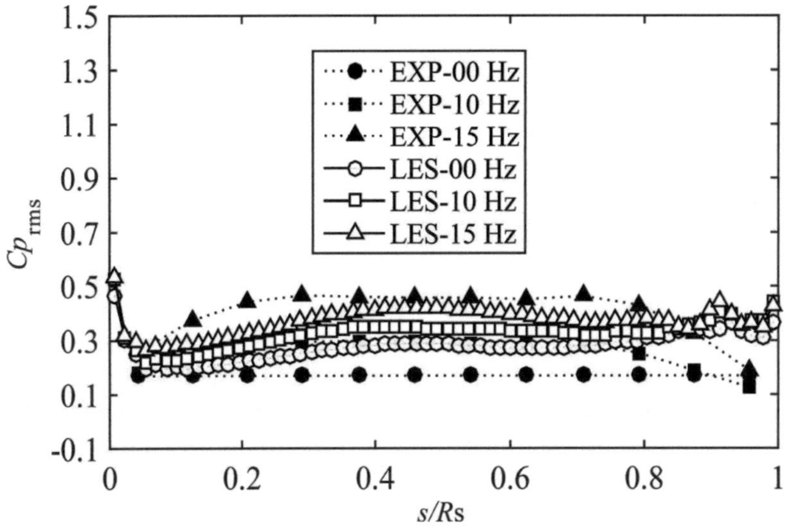

Figure 7. Comparisons for the distribution of the fluctuating wind pressure coefficients along the centerline between LES and wind tunnel experiment.

coefficients *Cp_mean* near the rooftop increase in magnitude and the rms fluctuating wind pressure coefficients *Cp_rms* generally increase, as the forced-vibration frequency increases. Furthermore, the variation is significant near the position of the greatest forced-vibration amplitude. These results indicate that the wind pressure field around the vibrating roof is strongly influenced by the vibration of the roof.

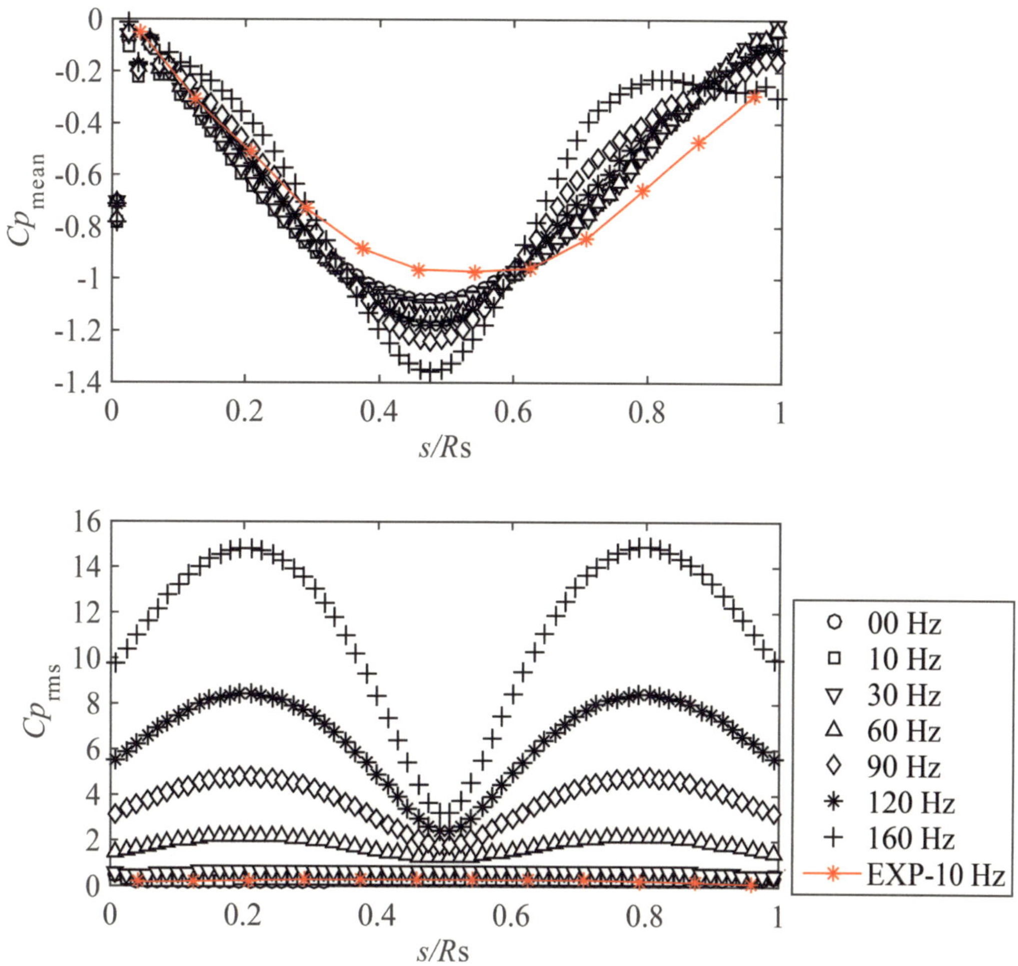

Figure 8. Variation of mean and fluctuating wind pressure coefficients with forced vibration frequency (r/L = 0.15).

Figure 9 shows the variations of mean and rms fluctuating wind pressure coefficients with the rise/span ratio. It can be seen that the *Cp_mean* changes from negative to positive at the leading edge of the roof as the rise/span ratio increases. Furthermore, the negative peak value increases in magnitude with an increase in rise/span ratio. The value of *Cp_rms* increases with an increase in rise/span ratio at the middle part of the roof. However, the effect of r/L on *Cp_rms* is less significant than on *Cp_mean*.

4.3. Discussion flow field around the roof

The roof configurations at several steps (phases) during one period of vibration are shown in **Figure 10**. The deformation of the windward side is upward and becomes the greatest at step 2; and that of the leeward side is upward and becomes the greatest at step 4.

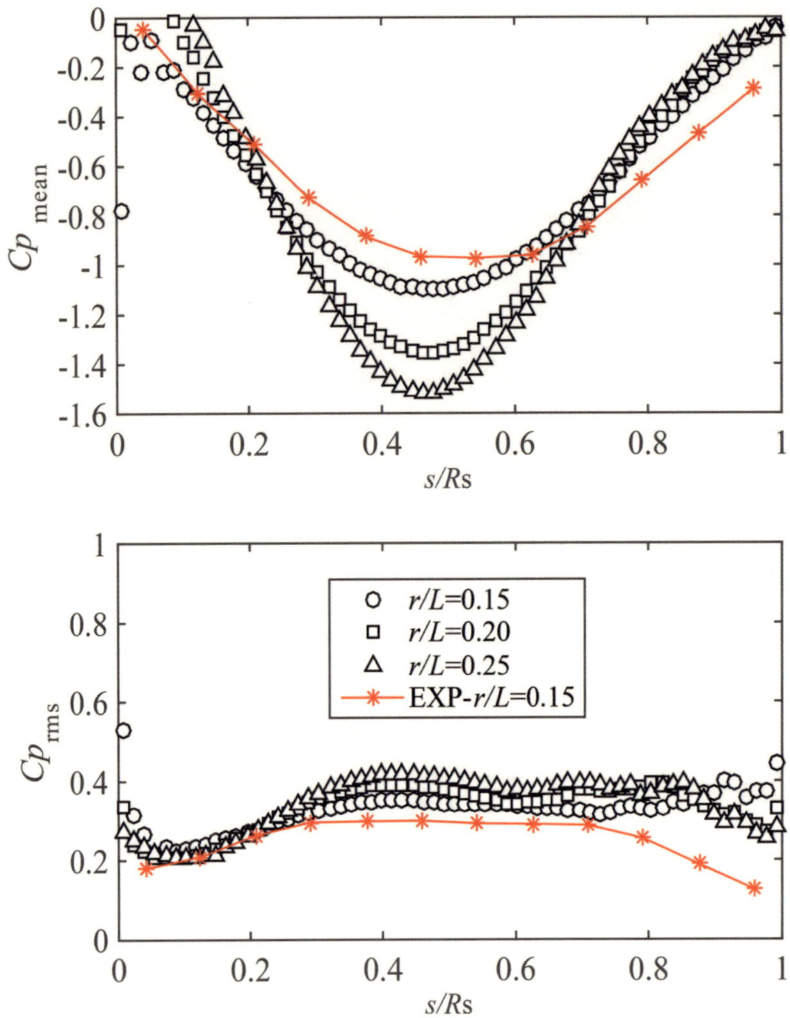

Figure 9. Variation of mean and fluctuating wind pressure coefficient with rise/span ratio (f_m = 10 Hz).

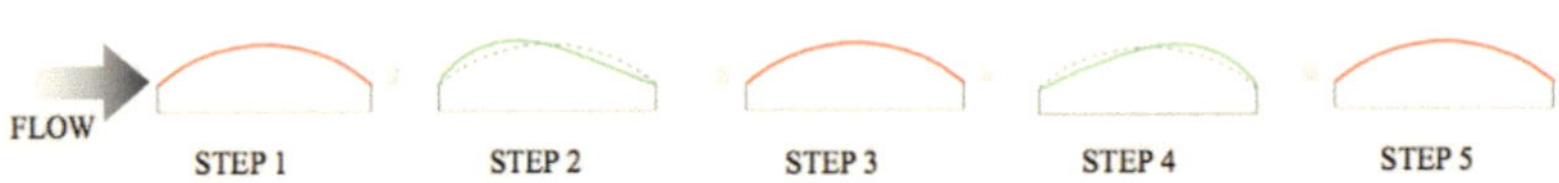

Figure 10. Roof configurations at several steps.

Figure 11 shows representative flow fields around a stationary or vibrating roof at a frequency of 10 or 20 Hz. It can be seen that the wind speed increases near the roof regardless of the roof's vibration. In the case of a stationary roof (f_m = 0 Hz), the flow separates near the 3/4 position of the roof from the leading edge. On the other hand, in the case of a vibrating roof, the separated vortex seems smaller than that in the stationary roof case, which may be due to the vibration of the roof that restrains the separation of the vortex. In addition, the separated position at the

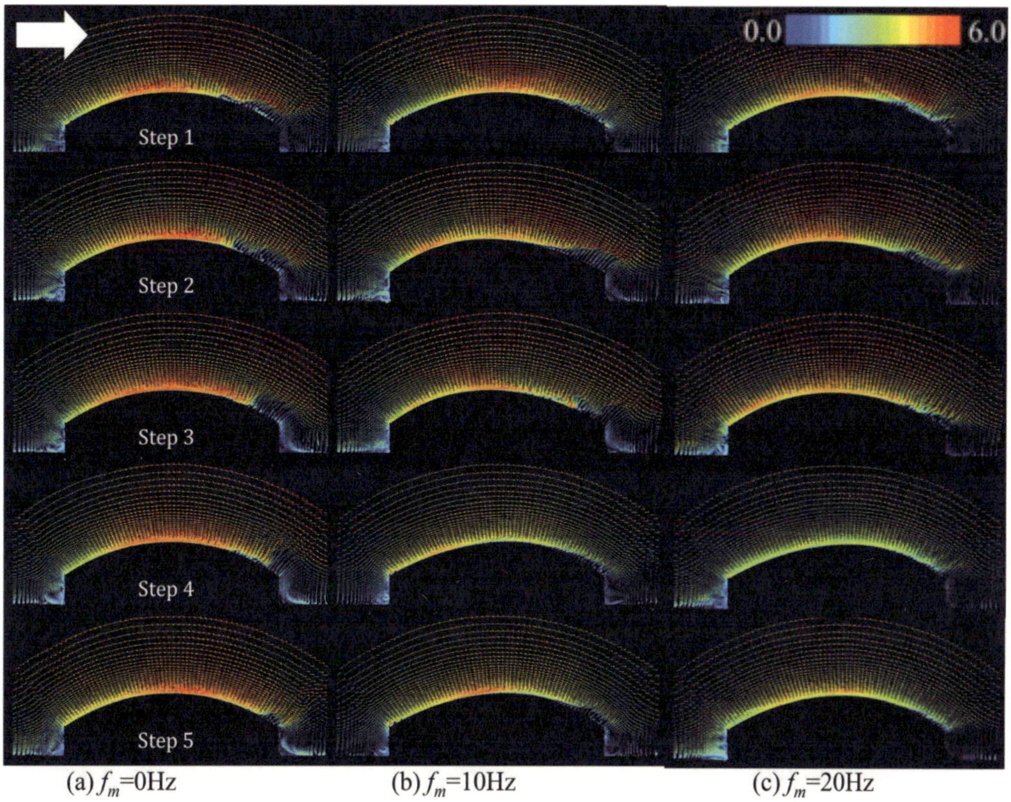

Figure 11. Flow fields around the roof for various forced vibration frequencies (r/L = 0.15); (a) 0 Hz; (b) 10 Hz; (c) 20 Hz.

rear of roof changes with the vibration of roof. The separated position is relatively forward when the roof is vibrated in step 1 to step 3, because the deformation at the windward side of roof makes the flow separated in advance. On the other hand, the separated position is relatively backward when the roof is vibrated in step 3 to step 5, as the result that the deformation at the leeward side of roof restrains the flow separated.

Figure 12 shows the effect of the rise/span ratio on the flow field around a vibrating roof at a forced vibration frequency of 20 Hz. It can be seen that the wind speed near the rooftop becomes higher, generating larger suction as the rise/span ratio increases. Therefore, the negative peak value of *Cp_mean* increases with an increase in rise/span ratio (see **Figure 8**). Furthermore, as the rise/span ratio increases, the vortex at the rearward of roof becomes larger. Flow fields around the roof for various rise/span ratios (f_m = 20 Hz); (a) r/L = 0.15; (b) r/L = 0.20; (c) r/L = 0.25.

4.4. Evaluation of unsteady aerodynamic forces

In this study, we use aerodynamic stiffness coefficient a_{Kj} and aerodynamic damping coefficient a_{Cj} to investigate the characteristics of unsteady aerodynamic forces acting on a vibrating long-span curved roof, which are given by the following equations [8]:

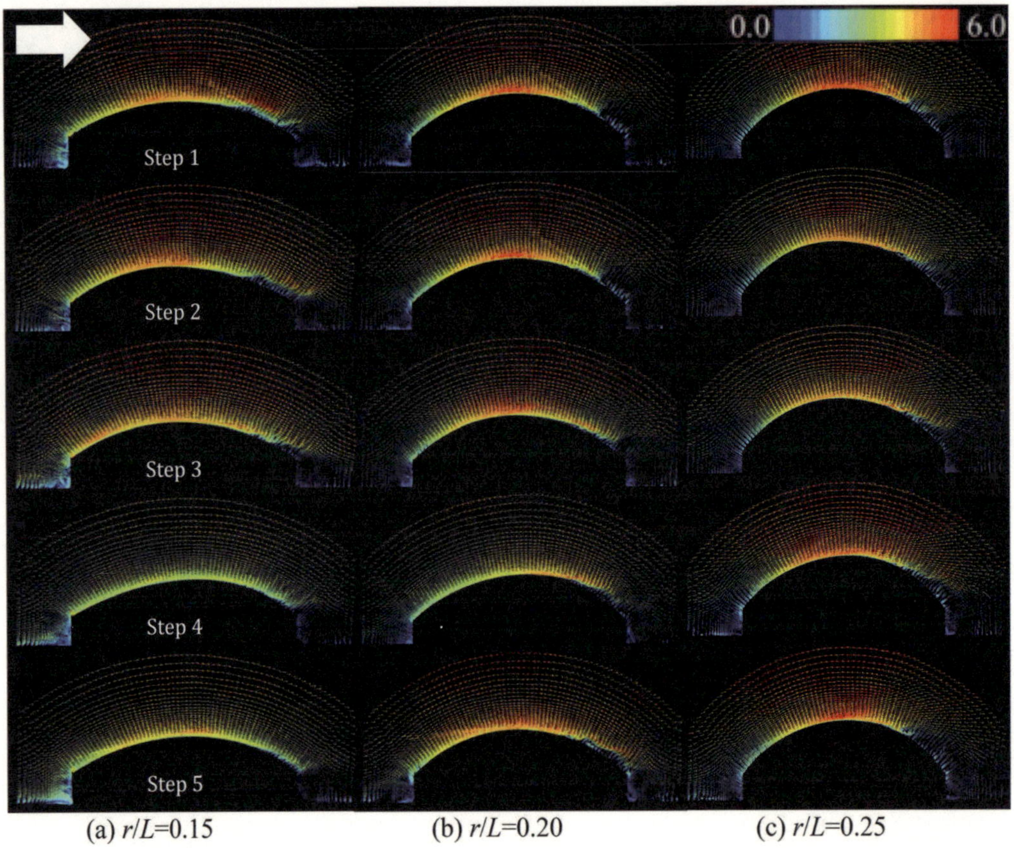

Figure 12. Flow fields around the roof for various rise/span ratios (f_m = 20 Hz); (a) r/L = 0.15; (b) r/L = 0.20; (c) r/L = 0.25.

$$a_{Kj} = \frac{F_{Rj}(f_m)}{q_H A_s(x_0/L)} = \frac{1}{q_H A_s(x_0/L)} \frac{1}{T} \int_{-T}^{T} F_j(t) \cos 2\pi f_m t \tag{14}$$

$$a_{Cj} = \frac{F_{Ij}(f_m)}{q_H A_s(x_0/L)} = \frac{1}{q_H A_s(x_0/L)} \frac{1}{T} \int_{-T}^{T} F_j(t) \sin 2\pi f_m t \tag{15}$$

where F_{Rj} is the in-phase component with the generalized displacement represented as the aerodynamic stiffness term, F_{Ij} is the in-phase component with velocity represented as the aerodynamic damping term, q_H = velocity pressure at the mean roof height H, A_s = roof area, x_0 = forced vibration amplitude, L = span of the roof, T = vibration period, f_m = forced vibration frequency, and f_m^* = reduced frequency of vibration, defined by f_mH/U_H, with U_H being the mean wind speed at the mean roof height H.

The generalized force F_j may be described in terms of the external and internal pressures p_e and p_i, as shown in Eq. 16,

$$F_j(t) = \int_0^{R_s} \left[p_{e_j}(s,t) - p_{i_j}(s,t) \right] \varphi_j(s) ds \tag{16}$$

http://dx.doi.org/10.5772/intechopen.70880

where R_s = total length of the vaulted roof. Internal pressure p_i is ignored in the present study, because the first anti-symmetric mode under consideration causes no change of internal volume. The model's vibration mode almost corresponded with the asymmetric sine mode, as shown in Eq. 17.

$$\varphi_j(s) = \sin 2\pi \frac{s}{R_s} \tag{17}$$

Figure 13 shows the aerodynamic stiffness and damping coefficients, a_K and a_C, obtained from the LES and the wind tunnel experiment, plotted as a function of the reduced frequency of vibration f_m^* ($f_m^* = f_m H/U_H$). The wind tunnel experiment was carried out in a limited range of f_m^*, while the LES was conducted over a wider range of f_m^*. It can be seen that the LES results are consistent with those of the wind tunnel experiment, which indicates that the LES model can be used for investigating the characteristics of unsteady aerodynamic forces. The aerodynamic stiffness coefficient a_K is generally positive and increases with an increase in f_m^*, which decreases the total stiffness of the system. On the other hand, the aerodynamic damping coefficient a_C is negative and increases in magnitude with an increase in f_m^*, resulting in an increase in the total damping of the system.

The distribution of aerodynamic stiffness and damping coefficients a_K and a_C with f_m^* for various rise/span ratios is shown in **Figure 14**. It can be seen that the values of a_K for r/L = 0.15, 0.20, and 0.25 are generally consistent with each other when f_m^*<0.4. However, when f_m^*>0.4, the value of a_K decreases with an increase in the rise/span ratio. Regarding the value of a_C, the results for various rise/span ratios are generally similar to each other. This figure indicates that the value of a_K is influenced by the rise/span ratio of a long-span vaulted roof. However, the effect of the rise/span ratio on the value of a_C is small.

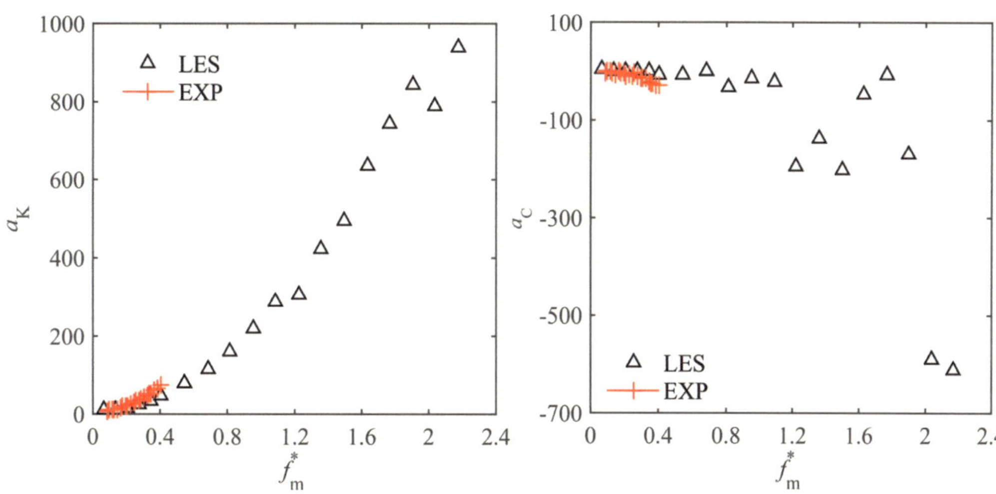

Figure 13. Comparisons of the LES and the wind tunnel experiment for the aerodynamic stiffness coefficient a_K and aerodynamic damping coefficient a_C versus reduced frequency of vibration f_m^*.

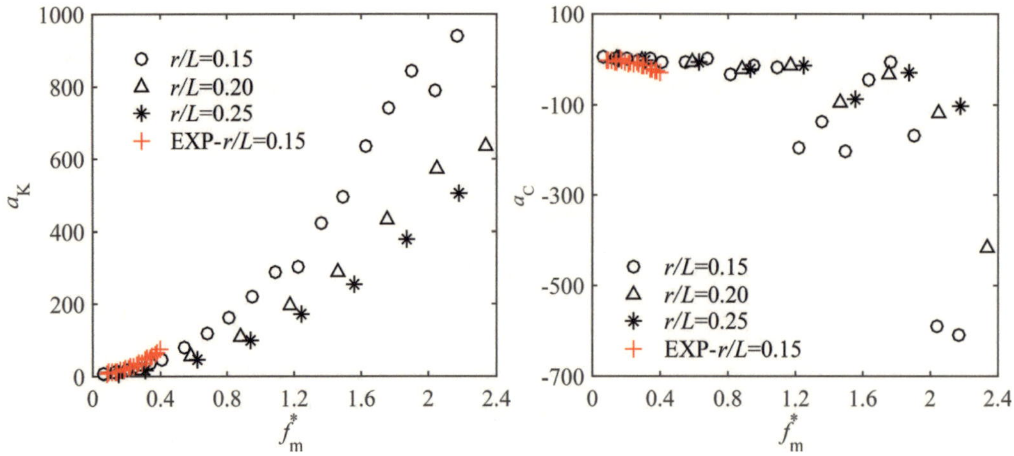

Figure 14. Aerodynamic stiffness and damping coefficients versus f_m^* for different rise/span ratio r/L.

5. Concluding remarks

The unsteady aerodynamic forces acting on the long-span curved roof have been investigated based on a numerical simulation (LES). A forced vibration test was carried out. First, the influence of a roof's vibration on the wind pressure was investigated. It is found that the wind pressure on a vibrating roof is strongly influenced by the roof's vibration. Furthermore, the flow field around a vibrating roof was also investigated. It is found that the vibration of the roof may restrain the separation of a vortex near the trailing edge of the roof. Finally, the characteristics of unsteady aerodynamic force acting on a long-span vaulted roof were evaluated. Both the wind tunnel experiment and CFD simulation show similar results for the variation of aerodynamic stiffness and damping coefficients a_K and a_C with reduced frequency of vibration f_m^*, which implies that the LES is effective to investigate the characteristics of unsteady aerodynamic forces. The aerodynamic stiffness coefficient is generally positive, which decreases the total stiffness of the system, resulting in aeroelastic instability of long-span vaulted roofs with lower stiffness. On the other hand, the aerodynamic damping coefficient is negative, which results in an increase in the total damping of the system, resulting in a decrease in the response of the roof.

Therefore, it is necessary to consider the effects of unsteady aerodynamic forces in the wind-resistant design of long-span curved roof with lightweight and low stiffness for evaluating the response of the roof more reasonably.

Author details

Wei Ding

Address all correspondence to: dingwei1985@live.cn

School of Mechanics and Civil Engineering, China University of Mining and Technology, Xuzhou, China

References

[1] Blackmore PA, Tsokri E. Wind loads on curved roof. Journal of Wind Engineering and Industrial Aerodynamics. 2006;**94**(11):833-844

[2] Chen FB, Li QS, Wu JR. Wind effects on a long-span beam string roof structure: Wind tunnel test, field measurement and numerical analysis. Journal of Wind Engineering and Industrial Aerodynamics. 2011;**67**(10):1591-1604

[3] Natalini MB, Morel C, Natalini B. Mean loads on vaulted canopy roof. Journal of Wind Engineering and Industrial Aerodynamics. 2013;**119**:102-113

[4] Wu Y, Chen ZQ, Sun XY. Research on the wind-induced aeroelastic response of closed-type saddleshaped tensioned membrane models. Journal of Zhejiang University: SCIENCE A (Applied Physics & Engineering). 2014;**16**(8):656-668

[5] Yang QS, Wu Y, Zhu WL. Experimental study on interaction between membrane structures and wind environment. Earthquake Engineering and Engineering Vibration. 2010;**9**(4): 523-532

[6] Chen ZQ, Wu Y, Sun XY. Research on the added mass of open-type one-way tensioned membrane structure in uniform flow. Journal of Wind Engineering and Industrial Aerodynamics. 2015;**137**:67-77

[7] Uematsu Y, Uchiyama K. Wind-induced dynamic behaviour of suspended roofs. The Technology Reports of the Tohoku University. 1982;**47**:243-261

[8] Daw DJ, Davenport AG. Aerodynamic damping and stiffness of a semi-circular roof in turbulent wind. Journal of Wind Engineering and Industrial Aerodynamics. 1989;**32**(1–2): 83-92

[9] Ohkuma T, Marukawa H. Mechanism of aeroelastically unstable vibration of large span roof. Wind Engineers, JAWE. 1990;**42**:35-42

[10] Ding W, Uematsu Y, Nakamura M, Tanaka S. Unsteady aerodynamic forces on a vibrating longspan curved roof. Wind and Structures. 2013;**19**(6):649-663

[11] Nozu T, Tamura T. Generation of unsteady wind data in boundary layers and its turbulence structures. In: The 15th National Symposium on Wind Engineering. 1998

[12] Lund TS, Wu X, Squires KD. Generation of turbulent inflow data for spatially-developing boundary layer simulation. Journal of Computational Physics. 1998;**140**(2):233-258

[13] Kataoka H, Mizuno M. Numerical flow computation around 3D square cylinder using inflow turbulence. Journal of Architecture Planning Environment Engineering. 1999;**523**: 71-77

[14] Nozawa K, Tamura T. Large eddy simulation of a turbulent boundary layer over a rough ground surface and evaluation of its fluctuating velocity profile. Journal of Structure and Construction Engineering. 2001;**541**:87-94

[15] Chikamatsu A, Nozawa K, Tamura T. Large eddy simulation of turbulent flows around a cube in an imitated atmospheric boundary layer. In: The 17th National Symposium on Wind Engineering; Tokyo, Japan. 2002

[16] Kondo K, Murakami S, Mochida A. Generation of velocity fluctuations for inflow boundary condition of LES. Journal of Wind Engineering and Industrial Aerodynamics. 1997;**67** & **68**:51-64

[17] Huang SH, Li QS, Wu JR. A general inflow turbulence generator for large eddy simulation. Journal of Wind Engineering and Industrial Aerodynamics. 2010;**98**:600-617

[18] Lu CL, Li QS, Huang SH, et al. Large eddy simulation of wind effects on a long-span complex roof structure. Journal of Wind Engineering and Industrial Aerodynamics. 2012; **100**(1):1-18

[19] Yan BW, Li QS. Inflow turbulence generation methods with large eddy simulation for wind effects on tall buildings. Computers and Fluids. 2015;**116**:158-175

Chapter 7

Wake Topology and Aerodynamic Performance of Heaving Wings

Joel E. Guerrero

Additional information is available at the end of the chapter

http://dx.doi.org/10.5772/intechopen.71517

Abstract

Simulating the three-dimensional flow features generated by heaving wings constitutes a great challenge due to the computational effort required to compute the complex three-dimensional flow produced as a function of the kinematics parameters, wing geometry, and Reynolds number. Hereafter, we study the wake topology generated by oscillating rigid wings and the validity of the Strouhal number as the fundamental parameters used to assess the aerodynamic performance of heaving wings. The unsteady laminar incompressible Navier-Stokes equations are solved on moving overlapping structured grids using a second-order accurate in space and time finite-difference numerical method. The numerical simulations are performed at a Reynolds number of Re = 250 and at different values of Strouhal number and heaving frequency.

Keywords: heaving wings, aerodynamic performance, wake topology, Strouhal number, overlapping grids

1. Introduction

Flapping wings for flying and oscillating fins for swimming stand out as one of the most complex yet widespread propulsion methods found in nature. Natural flyers and swimmers (which have evolved over millions of years) represent illuminating examples of biokinetics, unsteady aerodynamics, high maneuverability, endurance, and large aero/hydrodynamics efficiency.

In the field of flapping flight, biologist, zoologist, and engineers are sharing findings and conducting research together. From the point of view of a biologist or zoologist, studying flapping flight in nature is of great importance for understanding the biology, allometry, flight patterns, flight skills, and the migratory habits of avian life. From an engineering point of view, the main reason for studying flapping flight is the use of animal locomotion as inspiration for improving existing applications or developing new technologies by just mimicking nature evolutionary-optimization process (biomimetics). Such applications may include drag reduction, noise reduction, and flow control by using feather-like structures [1] and flippers tubercles [2]; the development of new propulsion/lift generation systems for micro-air-vehicles (MAVs), nano-air-vehicles (NAVs), and autonomous-underwater-vehicles (AUVs) is inspired by flapping wings or oscillating fins [3–8], energy harvesting applications [9], and even robotic extraterrestrial exploring missions [10].

An important aspect of flying using flapping wings or swimming by using oscillating fins or fanning is the ability to generate thrust with relatively high propulsive efficiency. Early attempts at building fish-inspired mechanisms achieved disappointingly low propulsive efficiencies [11]. It was only through a deeper understanding of the vorticity and wake produced by swimming animals, significant progress was achieved [12].

Many researchers [13–15] have found that flying and swimming animals cruise in a narrow range of Strouhal numbers (between 0.2 and 0.4), corresponding to a regime of vortex growth and shedding in which the propulsion efficiency peaks. The Strouhal number St is a dimensionless parameter defined as,

$$St = \frac{fh}{U} \tag{1}$$

where f is the flapping frequency, h is the peak to peak amplitude of the flapping stroke, and U is the forward velocity. This definition describes a ratio between the oscillating speed (fh) and the forward speed. Another dimensionless parameter that characterizes the aero/hydrodynamic performance and wake signature of flying and swimming animals is the reduced frequency k, which is a measure of the residence time of a vortex (or a particle) convecting over the wing/fin chord compared to the period of motion and is defined as,

$$k = \frac{\pi fc}{U} \tag{2}$$

Hence, it becomes evident that gaining a better understanding of the wing/fin motion parameters driving forces generation, vortices generation and shedding, the manner in which the vortices interact with the moving surfaces and themselves, and how they contribute to lift and propulsion would aid in better understanding the propulsion mechanism of birds, insects, and fishes, independently of their possible practical applications.

In the current numerical study, we aim at performing a comprehensive analysis of the wake signature and aerodynamic performance of finite-span rigid wings undergoing pure heaving motion. The laminar incompressible Navier-Stokes equations are numerically approximated, and all unsteady, viscous, and three-dimensional effects are solved. The simulations are conducted for Strouhal numbers values between $0.15 \leq St \leq 0.5$, and for two different reduced frequency values, one corresponding to high frequency and the other one to low frequency, this was done to study leading edge vortex shedding dependency.

The remainder of this paper is organized as follows. In Section 2, we give a brief description of the numerical method and gridding methodology. In Section 3, we present a description of the computational domain, case setup, and heaving kinematics. In Section 4, we present a short discussion of the quantitative and qualitative results obtained from a grid dependence study. In Section 5, we present a detailed discussion of the results. Finally, in Section 6, we present the conclusions and future developments.

2. Numerical method

Hereafter, we summarize the numerical method used to solve the governing equations on structured overlapping grids. For a complete description of the numerical method and gridding methodology, the interested reader should refer to the papers by Henshaw [16], Henshaw and Petersson [17], and Chesshire and Henshaw [18].

In primitive variables (u, v, w, p), the governing equations of the initial-boundary-value problem (IBVP) for the laminar incompressible Navier-Stokes equations can be written as

$$\frac{\partial \mathrm{u}}{\partial t} + \mathbf{u}\cdot\nabla\mathbf{u} = \frac{-\nabla p}{\rho} + \nu\nabla^2\mathbf{u} \qquad \mathit{for} \quad \mathbf{x}\in\mathcal{D}, \qquad t\geq 0, \tag{3}$$

$$\nabla\cdot\mathbf{u} = 0 \qquad \mathit{for} \quad \mathbf{x}\in\mathcal{D}, \qquad t\geq 0, \tag{4}$$

with the following boundary conditions and initial conditions,

$$B(\mathbf{u}, p) = \mathbf{g} \qquad \mathit{for} \quad \mathbf{x}\in\partial\mathcal{D}, \qquad t\geq 0, \tag{5}$$

$$\mathring{\mathbf{Q}}(x, 0) = \mathbf{q}_0(x) \qquad \mathit{for} \quad \mathbf{x}\in\mathcal{D}, \qquad t = 0. \tag{6}$$

In this IBVP, the vector $\mathbf{x} = (x, y, z)$ contains the Cartesian coordinates in physical space $\mathcal{P} = \mathcal{P}(x, y, z, t)$, $\mathcal{D}$ is a bounded domain in $\mathcal{P}\in\mathfrak{R}^{\mathbb{N}}$ (where $\mathbb{N}$ is the number of space dimensions), $\partial\mathcal{D}$ are the boundaries of the bounded domain $\mathcal{D}$, t is the physical time, $\boldsymbol{u}$ is a vector containing the velocity components (u, v, w), the scalar p is the pressure, and the constants ν and ρ are the kinematic viscosity and density. In Eq. (5), B is a boundary operator (that can leave to a Dirichlet or Neumann boundary condition), and $\mathbf{g}$ is the boundary condition input value. In Eq. (6), $\mathring{\mathbf{Q}}$ is the initial condition, and $\mathbf{q}_0$ is the input value of the initial conditions (which can be a uniform or nonuniform field).

An alternative formulation of the system of Eqs. (3)–(6), called the velocity-pressure formulation, can be written as follows,

$$\frac{\partial \mathbf{u}}{\partial t} + \mathbf{u}\cdot\nabla\mathbf{u} = \frac{-\nabla p}{\rho} + \nu\nabla^2\mathbf{u} \quad \mathit{for} \quad \mathbf{x}\in\mathcal{D}, \quad t\geq 0, \tag{7}$$

$$\frac{\nabla^2 p}{\rho} + \nabla u\cdot\mathbf{u}_x + \nabla v\cdot\mathbf{u}_y + \nabla w\cdot\mathbf{u}_z = 0 \quad \mathit{for} \quad \mathbf{x}\in\mathcal{D}, \quad t\geq 0, \tag{8}$$

with the following boundary and initial conditions,

$$B(\mathbf{u},p)=g \qquad \textit{for} \quad \mathbf{x}\in\partial\mathcal{D}, \quad t\geq 0, \tag{9}$$

$$\nabla\cdot\mathbf{u}=0 \qquad \textit{for} \quad \mathbf{x}\in\partial\mathcal{D}, \quad t\geq 0, \tag{10}$$

$$\mathring{\mathbf{Q}}(\mathbf{x},0)=q_0(\mathbf{x}) \qquad \textit{for} \quad \mathbf{x}\in\mathcal{D}, \quad t=0. \tag{11}$$

The system of Eqs. (7)–(11) is equivalent to the original formulation (Eqs. (3)–(6) [16, 19, 20]) and is the form of the equations that will be discretized. Hence, we look for an approximate numerical solution of Eqs. (7) and (8), in a given domain $\mathcal{D}$, with prescribed boundary conditions $\partial\mathcal{D}$ and given initial conditions $\mathring{\mathbf{Q}}$ (Eqs. (9)–(11)). Eqs. (7)–(11) are solved in logically rectangular grids in the transformed computational space (the interested reader should refer to the following papers for a detailed derivation [16, 18, 21, 22]), using second-order centered finite-difference approximations on structured overlapping grids.

The structured overlapping grids method consists in generating a set of conforming structured component grids $\mathcal{G}_g$ that completely cover the domain $\mathcal{D}$ that is being modeled in physical space $\mathcal{P}$ and overlap where they meet. In this newly generated overlapping grid system $\mathbb{G}$, domain connectivity is obtained through proper interpolation in the overlapping areas, and in the case of moving bodies, grid connectivity information is recomputed at each time step.

As the problem of heaving wings is implicitly a problem with moving bodies, we need to solve the governing equations in a frame that moves with the component grids. For moving overlapping grids, Eqs. (7) and (8) can be expressed in a reference frame moving with the component grids as follows,

$$\frac{\partial\mathbf{u}}{\partial t}+\left[\left(\mathbf{u}-\dot{\mathbf{G}}\right)\cdot\nabla\right]\mathbf{u}=\frac{-\nabla p}{\rho}+\nu\nabla^2\mathbf{u} \quad \textit{for} \quad \mathbf{x}\in\mathcal{D}, \quad t\geq 0, \tag{12}$$

$$\frac{\nabla^2 p}{\rho}+\nabla u\cdot\mathbf{u}_x+\nabla v\cdot\mathbf{u}_y+\nabla w\cdot\mathbf{u}_z=0 \quad \textit{for} \quad \mathbf{x}\in\mathcal{D}, \quad t\geq 0, \tag{13}$$

where $\dot{\mathbf{G}}$ represents the velocity of the components grids. It is worth noting that the new governing equations (Eqs. (12) and (13)) must be accompanied by proper boundary conditions. For a moving body with a corresponding moving no-slip wall (e.g., a heaving wing), we should impose the velocity at the wall as follows,

$$\mathbf{u}(\mathbf{x}_{wall},t)=\dot{\mathbf{G}}(\mathbf{x}_{wall},t), \quad \textit{where} \quad \mathbf{x}_{wall}\in\partial\mathcal{D}_{wall}(t). \tag{14}$$

After spatial discretization of the governing equations, we can proceed with the temporal discretization. By proceeding in this way, we are using the method of lines (MOL) [23, 24]. The main advantage of the MOL method is that it allows to select numerical approximations of different accuracy for the spatial and temporal terms. Each term can be treated differently to yield different accuracies.

At this point, the discretized equations can be integrated in time by using any time discretization method. In this work, we used a semi-implicit multistep method, where the viscous terms are approximated using the Crank-Nicolson scheme, and the convective terms are approximated using the Adams-Bashforth/Adams-Moulton predictor-corrector approach (where the velocity is advanced in time by using a second-order Adams-Bahforth predictor step, followed by a second-order Adams-Moulton corrector step). We chose to implicitly treat the viscous terms because if they were treated explicitly, we could have a severe time step restriction proportional to the spatial discretization squared. The solution method previously described yields to a second-order accurate in space and time numerical scheme on structured overlapping grids.

To assemble the overlapping grid system and solve the laminar incompressible Navier-Stokes equations in their velocity-pressure formulation, the overture framework was used (http://www.overtureframework.org/). The large sparse system of linear algebraic equations arising from the discretization of the laminar incompressible Navier-Stokes equations is solved using the PETSc library (http://www.mcs.anl.gov/petsc/), which was interfaced with overture. The system of equations is then solved using a Newton-Krylov iterative method, in combination with a suitable preconditioner and in parallel computational architectures.

3. Geometry, boundary conditions, initial conditions, and wing kinematics

In **Figure 1**, we present an illustration of the overlapping grid system layout used to conduct this parametric study. In the figure, c is the wing's chord, and h is the heaving amplitude of the heaving wing. The background grid (BG) extends $4.0 \times c$ away from the wing's leading edge (LE), $10.0 \times c$ away from the wing's trailing edge (TE), $2.0 \times c$ away from the left and right wing's tips (LH-WT and RH-WT, respectively), and $4.0 \times c+h$ away from the point of maximum thickness of the upper and lower surfaces. This overlapping grid system layout corresponds to the instant, when the wing is in the mid position of the heaving cycle (as illustrated in **Figure 1**). In this manuscript, all the base units are expressed in the international system.

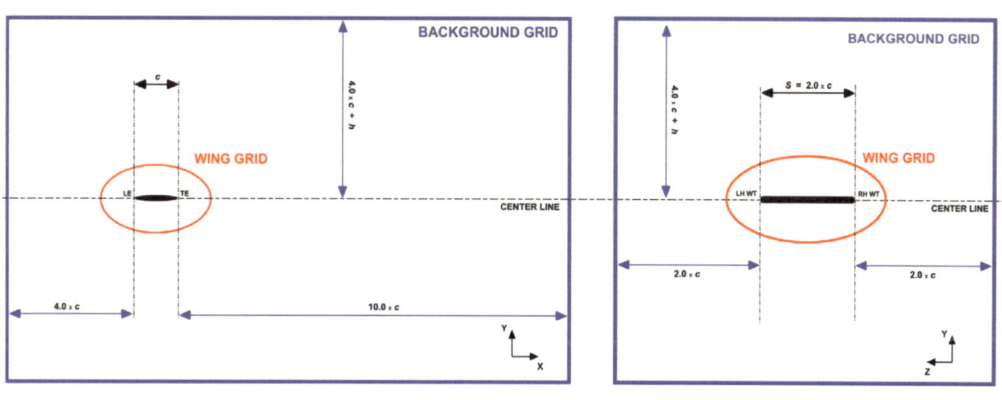

Figure 1. Left: computational domain layout in the xy plane. Right: computational domain layout in the zy plane. The figure is not to scale.

In all the cases studied, a rectangular wing with an aspect ratio AR equal to 2 was used. The cross-section of the wing is an ellipse, with a corresponding major axis $a = 0.25$ and a minor axis $b = 0.025$. Therefore, the wing's chord c is equal to $2 \times a = 0.5$.

The initial conditions used for all the heaving wings simulations are those of a fully converged solution of the corresponding fixed wing case. In **Figure 1** (left), the left boundary of the BG corresponds to an inflow boundary condition ($\mathbf{u} = (1.0, 0.0, 0.0)$, $\partial_{\hat{\mathbf{n}}} p = 0$), and the top, bottom, and right boundaries of the BG are outflow boundaries (velocity extrapolated from the interior points). In **Figure 1** (right), all the boundaries of the BG correspond to outflow conditions. On the wing surface (which is a moving body), we impose a no-slip boundary condition for moving walls ($\mathbf{u} = \dot{\mathbf{G}}_x, \dot{\mathbf{G}}_y, \dot{\mathbf{G}}_z$). The rest of the boundaries is interpolation boundaries, where we used a nonconservative Lagrange interpolation scheme. The Reynolds number (defined as $\mathrm{Re} = U \times c/\nu$) is equal to 250 for all the simulations.

In all the simulations conducted, we assumed that the wing is undergoing pure heaving motion, wherein the wing cross-section heaves in the vertical direction (or y axis in **Figure 1**) and according to the following function,

$$y(t) = h \times \sin\left(2 \times \pi \times f \times t + \phi\right) \tag{15}$$

where $y(t)$ is the heaving motion (and is defined positive upwards), h is the heaving amplitude, f is the heaving oscillating frequency, ϕ is the phase angle of the heaving motion (0 in this case), and t is the physical time.

4. Grid dependence study and vortical structures visualization

When conducting numerical simulations, it is well known that the grid resolution can affect the results from a quantitative and qualitative point of view. In this section, we present the outcome of the grid dependence study used to determine the best overlapping grid system $\mathbb{G}$ in terms of computing time, stability, and accuracy of the solution. To conduct this study, we used the grid convergence index method or GCI, as described by Roache in [25, 26].

During this study, fixed and moving wings were considered, but for simplicity, we will only present the results related to pure heaving motion (where the uncertainties are higher). Several simulations were run at a Strouhal number $St = 0.3$ and at a reduced frequency $k = 1.570795$, and unsteadiness was observed to disappear typically after 4 to 5 cycles of wing heaving motion, and further calculations show negligible nonperiodicity. Each simulation was checked for acceptable iterative convergence.

The different grid sizes, layouts, and mesh stretching ratios considered during this study are shown in **Table 1**. In **Table 1**, the grid spacing ratio (GSR) is the refinement ratio from the finer grid to the coarser grid and is expressed in reference to the position of the first node normal to the wing surface (1NW). Therefore, the grid spacing refinement ratio r is equal to 2. In **Figure 2**, we illustrate a typical overlapping grid system $\mathbb{G}$ used in this study. In **Figure 2**, the background grid is shown in images a, b, and c (rectangular grid). The wing grid is made up of

Grid $\mathbb{G}_g$	BG	WG-CS	WG-TS	GSR	1NW
$\mathbb{G}_1$	161 × 121 × 101	221 × 121 × 41	81 × 61 × 41	1	0:001 × 2c
$\mathbb{G}_2$	161 × 121 × 101	201 × 101 × 31	61 × 51 × 31	2	0:002 × 2c
$\mathbb{G}_3$	161 × 121 × 101	161 × 81 × 31	51 × 41 × 31	4	0:004 × 2c

BG stands for background grid, WG-CS stands for wing grid center-section, WG-TS stands for wing grid tip-section (only one tip-section), GSR stands for grid refinement spacing ratio, and 1NW stands for position of the first node normal to the wall.

Table 1. Grid dimensions used for the grid dependence study.

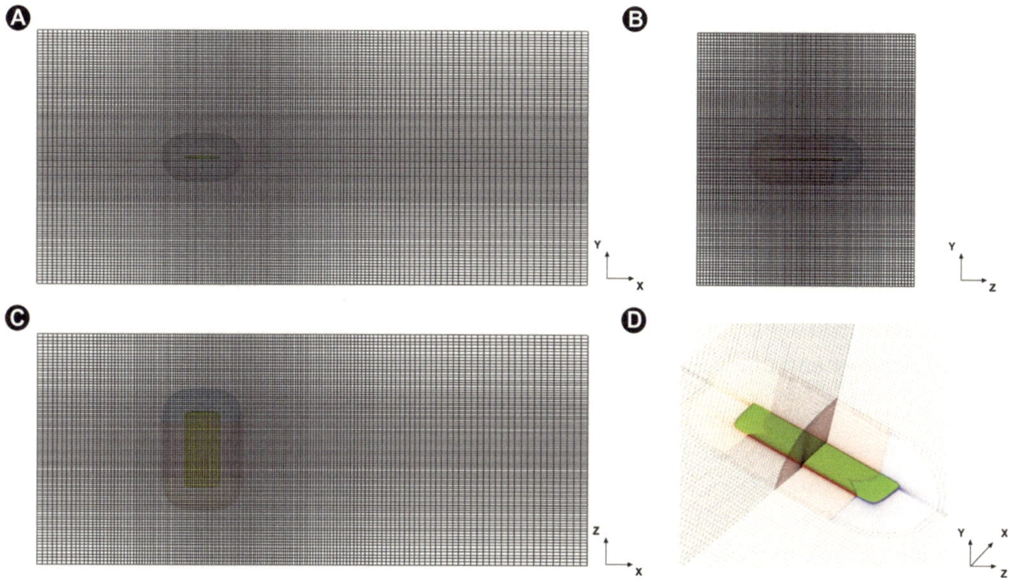

Figure 2. Typical overlapping grid system $\mathbb{G}$ used in this study. (A) Side view. (B) Front view. (C) Bottom view. (D) Perspective view.

three component grids (image d), namely, center section, left wing grid tip-section, and right wing grid tip-section.

Since in the study of heaving wings propulsion, the main task is thrust production, and it is more convenient to think in terms of thrust force T instead of drag force D. The thrust force is equal in magnitude but opposite in direction to the drag force ($T = -D$). To quantify the unsteady aerodynamics performance, we computed the lift coefficient c_l, the drag coefficient c_d, and thrust coefficient c_t as follows,

$$c_l = \frac{L}{\frac{1}{2}\rho U^2 A}, \quad c_d = \frac{D}{\frac{1}{2}\rho U^2 A}, \quad c_t = \frac{T}{\frac{1}{2}\rho U^2 A} \tag{16}$$

In Eq. (16), the lift force L and the thrust force T (where $T = -D$) are computed by integrating the viscous and pressure forces over the wing surface.

Grid $\mathbb{G}g_g$	GSR	$\overline{c_d}$	c_l^{rms}
$\mathbb{G}_1$	1	0.057892	0.905822
$\mathbb{G}_2$	2	0.058476	0.897486
$\mathbb{G}_3$	4	0.060914	0.865326

Table 2. Observed values of $\overline{c_d}$ and c_l^{rms} for the grid dependence study.

In **Table 2**, we present the average drag coefficient $\overline{c_d}$ and the root mean square of the lift coefficient c_l^{rms} for the grid dependence study. These values were used to compute the observed order of convergence p_{obs}, the extrapolated value of the observed quantity at zero grid spacing $\chi_{spacing=0}$, the fine grid convergence index GCI_{12} and GCI_{23}, and the constancy of $GCI_{23} = r^{p_{obs}} \times GCI_{12}$.

Based on the outcome of the GCI study (refer to **Table 3**), $\overline{c_d}$ at zero grid spacing is estimated to be 0.057708, with an error band of 0.397203% for grid $\mathbb{G}_1$ and an error band of 1.641617% for grid $\mathbb{G}_2$, whereas c_l^{rms} at zero grid spacing is estimated to be 0.908738 with an error band of 0.402503% for grid $\mathbb{G}_1$ and an error band of 1.567203% for grid $\mathbb{G}_2$. The constancy of $GCI_{23} = r^{p_{obs}} \times GCI_{12}$ indicates that the results for grids $\mathbb{G}_1$ and $\mathbb{G}_2$ are within the asymptotic range of convergence. Finally, the observed order of convergence p_{obs} represents a direct indication of the accuracy of the numerical method. In this study, we obtained a value of p_{obs} close to 2 for both quantities of interest, which indicate that the method is second-order accurate in space and time.

To supplement the quantitative GCI study, we conducted an additional grid dependence study but from the qualitative point of view (vortical structures resolution on the wake of the wing). The qualitative results for grids $\mathbb{G}_1$ and $\mathbb{G}_2$ (refer to **Table 1**) are illustrated in **Figure 3**. In **Figure 3**, we use the iso-surfaces of vorticity magnitude ($|\omega|$-criterion), and the iso-surfaces of Q-criterion [27, 28], to capture the vortices and their corresponding cores. As depicted in this figure, there are no discernible differences between the solutions, indicating that both grids are adequate for vortical structures resolution. It can also be seen that the $|\omega|$-criterion, although capable of capturing the general vortical structures, has the disadvantage of also showing the shear layers near the wing surface and between the vortices. On the other hand, the Q-criterion shows the details of the vortical structures more clearly as it does not show the shear layers close to the wing (as illustrated in **Figure 4**). Apart from this, both methods provide nearly

Outcome	$\overline{c_d}$	c_l^{rms}
pobs	2.061650	1.947838
$\chi_{spacing=0}$	0.057708	0.908738
GCI_{12} (%)	0.397203	0.402503
GCI_{23} (%)	1.641617	1.567203
$(GCI_{23}/GCI_{12}) \times (1/r_{pobs})$	0.990012	1.009249

Table 3. GCI study results.

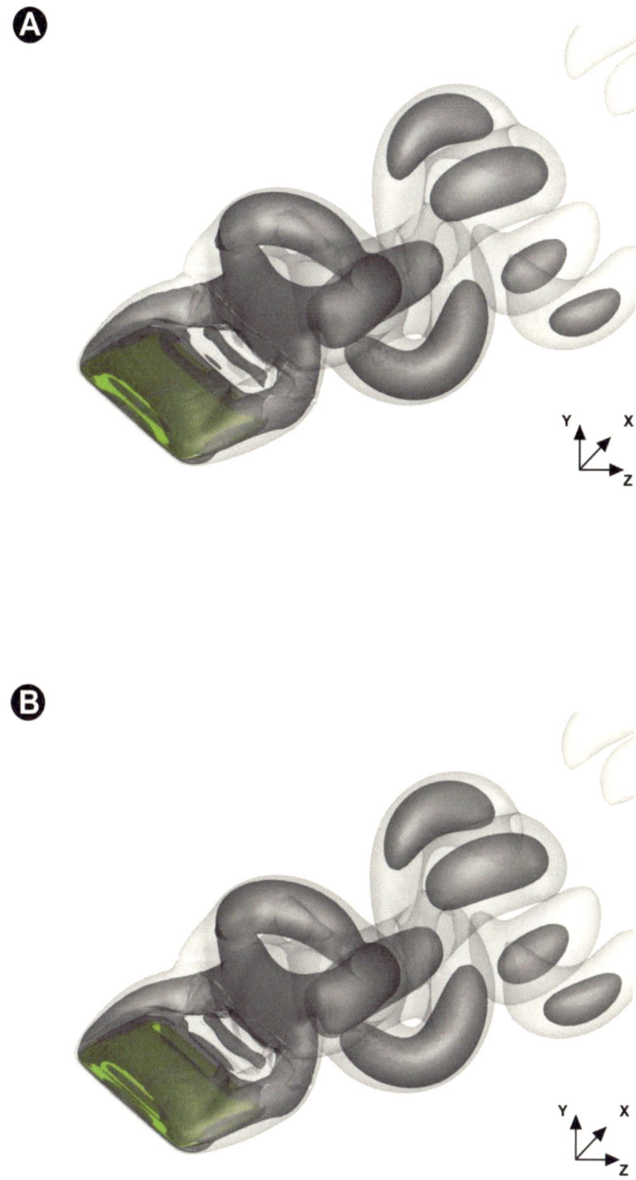

Figure 3. Vortex topology at the beginning of the upstroke ($t = 5.0$). Iso-surfaces of $|\omega|$-*criterion* are shown in light gray, and iso-surfaces of *Q-criterion* are shown in dark gray. Simulation parameters: $\mathrm{Re} = 250$, $St = 0.3$, and $k = 1.570795$. A corresponds to overlapping grid system $\mathbb{G}_1$ and B corresponds to overlapping grid system $\mathbb{G}_2$.

identical structures in the far wake. Based on these qualitative results, we chose the *Q*-criterion as the main criterion for wake topology characterization.

Summarizing the quantitative and qualitative results previously presented, we can conclude that the solutions obtained by using the overlapping grid systems $\mathbb{G}_1$ and $\mathbb{G}_2$ are grid independent. Taking into account the computational resources available, CPU time restrictions, and

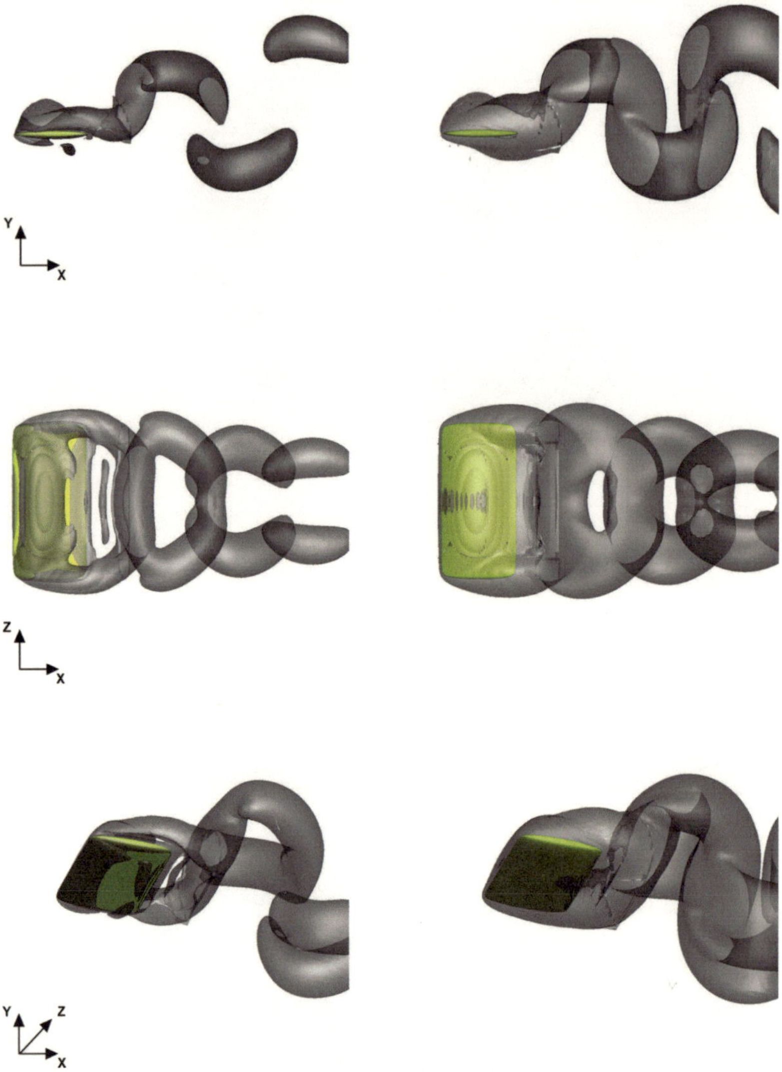

Figure 4. Shear layers close to the wing at the beginning of the upstroke ($t = 5.0$). Left column corresponds to iso-surfaces of *Q-criterion*, and the right column corresponds to iso-surfaces of $|\omega|$-*criterion*. Simulation parameters: Re = 250, $St = 0.3$, and $k = 1.570795$.

solution accuracy, $\mathbb{G}_2$ with a value of 1 NW equal to $0.001 \times 2c$ is used as the base grid to perform all further computations. In the case of a smaller or bigger computational domain, the grid dimensions are scaled in order to keep the same grid spacing as for this domain.

5. Simulation results

Hereafter, we carry out a comprehensive parametric study to assess the wake signature and aerodynamic performance of heaving rigid wings. In **Table 4**, we present the kinematics

http://dx.doi.org/10.5772/intechopen.71517

Case number	St	k	$\overline{c_t}$	$\widehat{c_l}$	Regime
3DH-1	0.15	1.570795	−0.122675	0.560613	Drag production
3DH-2	0.15	0.785397	−0.111949	0.422249	Drag production
3DH-3	0.20	1.570795	−0.106101	0.784628	Drag production
3DH-4	0.20	0.785397	−0.087877	0.597156	Drag production
3DH-5	0.25	1.570795	−0.086136	1.040502	Drag production
3DH-6	0.25	0.785397	−0.068118	0.798169	Drag production
3DH-7	0.30	1.570795	−0.058294	1.238780	Drag production
3DH-8	0.30	0.785397	−0.052475	1.024220	Drag production
3DH-9	0.35	1.570795	0.022962	1.651670	Thrust production
3DH-10	0.35	0.785397	−0.036191	1.257870	Drag production
3DH-11	0.40	1.570795	0.061370	2.030980	Thrust production
3DH-12	0.40	0.785397	0.008296	1.539740	Thrust production
3DH-13	0.45	1.570795	0.143680	2.417032	Thrust production
3DH-14	0.45	0.785397	0.010296	1.891345	Thrust production
3DH-15	0.50	1.570795	0.199366	2.944020	Thrust production
3DH-16	0.50	0.785397	0.024617	2.150710	Thrust production

Table 4. Simulation results for the pure heaving parametric study (positive $\overline{c_t}$ indicates thrust production whereas negative $\overline{c_t}$ indicates drag production).

parameters governing the heaving motion (described by Eq. (15)). In **Table 4**, we also present the quantitative results obtained, where $\overline{c_t}$ is the average thrust coefficient and $\widehat{c_l}$ is the maximum lift coefficient (which was measured during the downstroke). By inspecting these results, we observe that as we increase St and k, the values of $\overline{c_t}$ and $\widehat{c_l}$ also increase. This result is expected because as we increase St and k (therefore, the oscillating frequency and heaving amplitude), the vertical velocity of the wing is higher. Hence, the forces excerpted on the wing's surface are larger. We also observe two different behaviors of the aerodynamic forces for high and low reduced frequencies k values. Hence, it seems that for heaving wings, the oscillating frequency plays an important role in the vortex generation and shedding and, henceforth, on the aerodynamic forces. These frequency dependence observations are similar to those of Wang [29], Young and Lai [30], and Guerrero [31], but here, we extend them to three-dimensional cases.

In **Table 4**, we can read that for values of $St<0.30$, the heaving wing produces drag; for values of $0.30<St<0.35$, the heaving wing produces little or no drag (or thrust), whereas for values of $St>0.35$, the heaving wing produces thrust. These results suggest that there is a range of St values, where the heaving wing generates thrust. The results also point at the presence of a combination of St and k values, where propulsive efficiency peaks. In general, the results are inline with the hypothesis that flying and swimming animals cruise at Strouhal numbers corresponding to a regime of vortex growth and shedding in which propulsion efficiency is high.

To understand how the heaving wing generates thrust, we need to take a closer look at the vortex shedding process. In **Figures 5** and **6**, we illustrate the vortex topology for a heaving case in the thrust producing regime (case 3DH-11 in **Table 4**). These figures show that the downstream wake (or far field wake) of this case consists of two sets of doughnut-shaped vortex rings (VR) which convect at oblique angles about the centerline of the heaving motion. Thus, the flow induced by each vortex ring along its axis is expected to have a net streamwise component linked to thrust production. This net streamwise momentum excess in the wake is connected with the thrust production of heaving wings. The process by which the vortex rings are formed can be explained by examining the vortex formation and shedding close to the

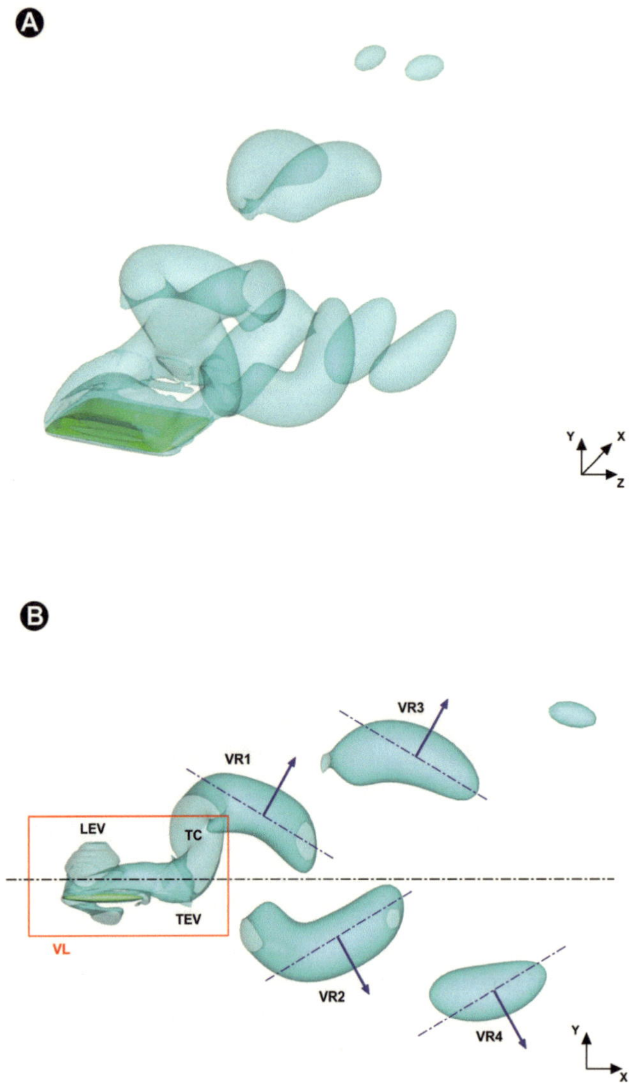

Figure 5. Vortex wake topology at the beginning of the upstroke for case 3DH-11 ($t = 5.0$). Heaving parameters: Re = 250, $St = 0.4$, and $k = 1.570795$ (thrust producing wake).

http://dx.doi.org/10.5772/intechopen.71517

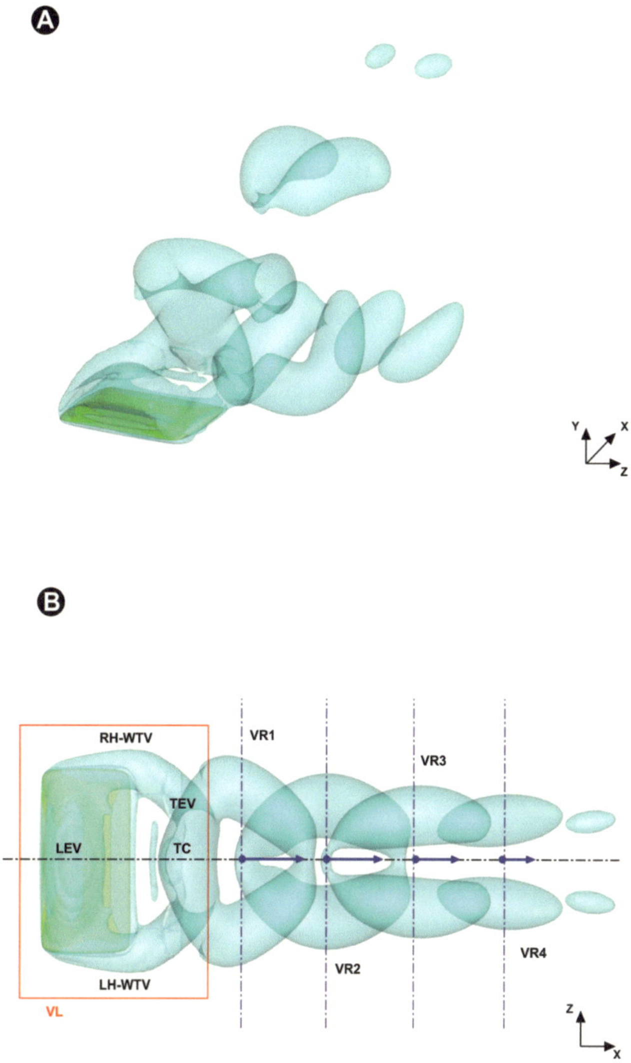

Figure 6. Vortex wake topology at the beginning of the upstroke for case 3DH-11 ($t = 5.0$). Heaving parameters: $Re = 250$, $St = 0.4$, and $k = 1.570795$ (thrust producing wake).

wing's surface. During the heaving motion and close to the wing, four vortices are formed; namely, one leading edge vortex (LEV), one trailing edge vortex (TEV), and two wing-tip vortices or WTV (one WTV on the left wing-tip or LH-WTV and another WTV on the right wing-tip or RH-WTV). These four vortices are all connected and form a closed vortex loop (VL). During the heaving motion and as this vortex loop is convected downstream, it disconnects from the wing, creating in this way the doughnut-shaped vortex rings. It is also of interest in the fact that each vortex loop has two sets of thin contrails (TC) that connect the VL to the VR generated in the previous stroke; these structures are segments of the wing-tip vortices, and, as the vortex rings are convected downstream, they become weaker and ultimately disappear (as in vortex ring VR2 in **Figure 5**). During a complete heaving cycle, two VRs are formed, one at the end of the upstroke and the other one at the end of the downstroke.

In **Figure 7**, we present the wake topology for a drag-producing case (which corresponds to case 3DH-3 in **Table 4**). It is clear from this figure that the wake topology is very different from the one of the thrust producing case. In this case, as the vortex loops are convected downstream, they do not morph into vortex rings. Instead, they keep their original shape, and as they are convected, they diffuse. We can also observe that the wake height is very compact, in comparison to that of the thrust production case (as depicted in **Figure 8**). Finally, notice how the flow induced by each vortex loop is inclined in the same direction of the wing's travel direction, resulting this in a momentum surfeit linked to drag production. The momentum

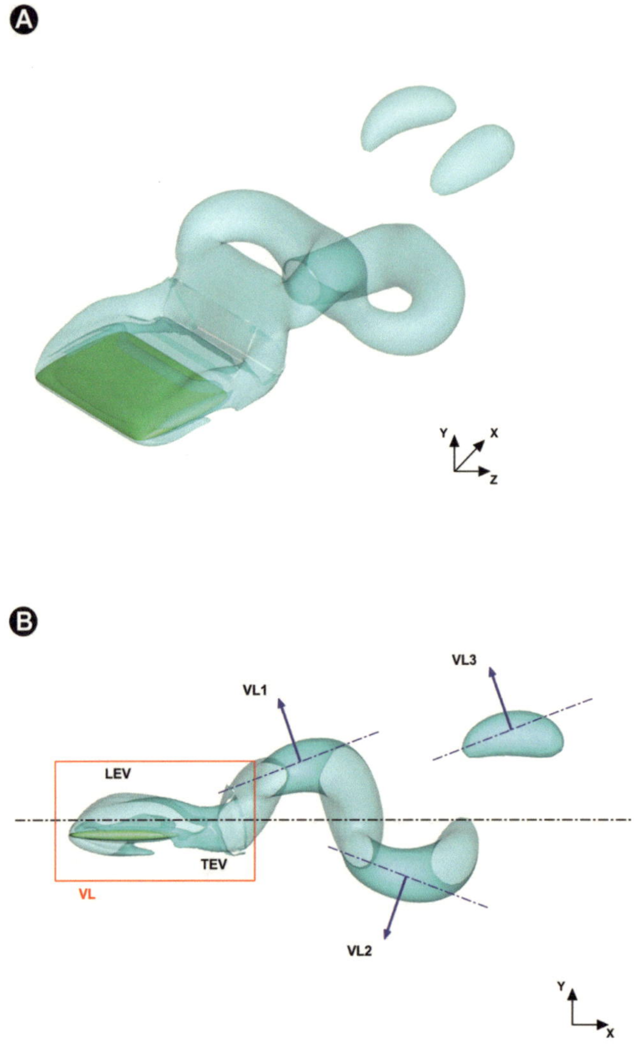

Figure 7. Vortex wake topology at the beginning of the upstroke for case 3DH-3 ($t=5.0$). Heaving parameters: Re = 250, $St=0.2$, and $k=1.570795$ (drag-producing wake).

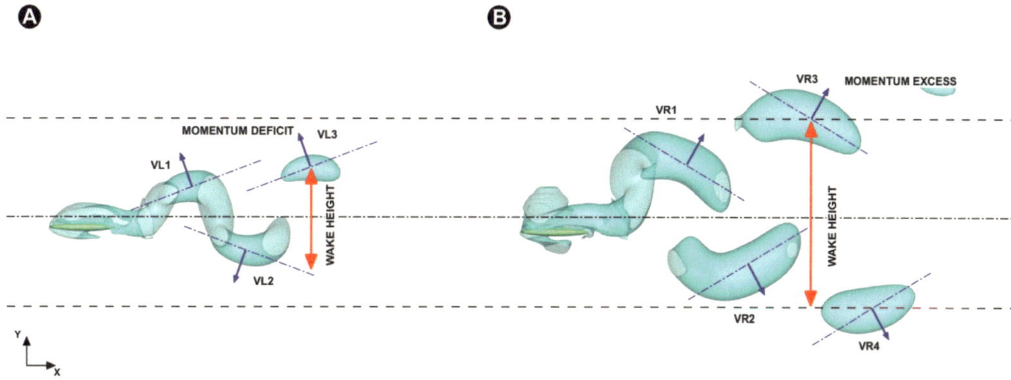

Figure 8. Wake comparison of a drag-producing case (A) and thrust producing case (B). The wake height was measured approximately at $3 \times c$ behind the trailing edge.

deficit and momentum excess scenarios can be better appreciated in **Figure 8**, where A corresponds to drag production (momentum deficit) and B corresponds to trust production (momentum excess).

Animations of several simulation cases are available at the author's web site: http://www.dicat.unige.it/guerrero/flapsim1.html (last accessed: Sep 2017).

6. Conclusions and perspectives

In this manuscript, we studied the unsteady aerodynamics of heaving rigid wings. The laminar incompressible Navier-Stokes equations were solved in their velocity-pressure formulation using a second-order accurate in space and time finite-difference numerical method, and to efficiently deal with moving bodies, we used overlapping structured grids. To study the dependence of the aerodynamic forces and wake topology on the wing kinematics, many simulations were conducted at different values of Strouhal number and at two reduced frequency values (low and high oscillating frequency).

The simulations show that the wake of thrust producing, rigid heaving wings is formed by two sets of interconnected vortex loops that slowly convert into vortex rings as they are convected downstream. It is also observed that the vortex rings are inclined with respect to the freestream flow, whereas for thrust producing configurations, the angle of inclination of the vortex rings is in the same direction of their travel, and for drag-producing configurations, the angle of inclination of the vortex rings is opposite to the direction of their travel. The presence of thin contrails that link the vortex loops is of interest; these structures are segments of the wing-tip vortices, and as the vortex loops are convected downstream, they become weaker and ultimately disappear. In general, the observed structures are qualitatively similar to those observed in the experiments by Parker et al. [32] and Von Ellenrieder et al. [33] and the numerical simulations of Dong et al. [34] and Blondeaux et al. [35].

From the force measurement study, two different behaviors were observed for the average thrust coefficient $\overline{c_t}$ and maximum lift coefficient $\widehat{c}_l$. It seems that the reduced frequency k (and hence the oscillating frequency) plays an important role in the vortex generation and shedding frequency, therefore, on the aerodynamic forces. Thus, LEV convection and separation introduce a frequency dependence into the results. This provides a mechanism of optimal selection of heaving frequency (in the sense of propulsive efficiency) as discussed by Wang [29], Guerrero [31], and Young and Lai [36]. It is worth mentioning that the results presented in the previous references were obtained for two-dimensional airfoils. The results presented in this manuscript extend these observations to three-dimensional wings.

Finally, for the limited range of St and k values studied and the simplified wing geometry and heaving kinematics covered in this study, all the qualitative and quantitative results presented are in close agreement with the experimental observations of Rohr and Fish [12], Triantafyllou et al. [13], Nudds et al. [14], Taylor et al. [15], and Parker et al. [32]; this supports the hypothesis that *"flying and swimming animals cruise at a Strouhal number tuned for high power efficiency"* [15].

The results presented in this manuscript are limited to laminar flow; nevertheless, they provide an excellent insight into the wake signature of the unsteady aerodynamics of heaving wings. We envisage to extend the current study to higher Reynolds numbers and turbulent cases and use more realistic wing geometries and kinematics. Finally, in this manuscript, we did not cover propulsive efficiency and optimal frequency selection, but we hope to address these issues in future studies.

Acknowledgements

The use of the computing resources at CINECA high performance computing center was possible thanks to the ISCRA grant 2016, project IsC45_DO4EnD. We also thank the DLTM (http://www.dltm.it) for letting us use their HPC facilities.

Author details

Joel E. Guerrero

Address all correspondence to: joel.guerrero@unige.it

University of Genoa, Genoa, Italy

References

[1] Favier J, Dauptain A, Basso D, Bottaro A. Passive separation control using a self-adaptive hairy coating. Journal of Fluid Mechanics. 2009;**627**:451-483

[2] Favier J, Pinelli A, Piomelli U. Control of the separated flow around an airfoil using a wavy leading edge inspired by humpback whale flippers. Comptes Rendus Mecanique. 2012;**340**:107-114

[3] Jones KD, Platzer MF. Experimental investigation of the aerodynamic characteristics of flapping-wing micro air vehicles. In: AIAA Paper 2003-0418-CP. 2003. pp. 1-11

[4] Singh SN, Simha A, Mittal T. AUV maneuvering by pectoral fins: Inverse control design based on CFD parametrization. IEEE Journal of Oceanic Engineering. 2004;**29**:777-785

[5] McMichael JM, Francis MS. Micro Air Vehicles toward a New Dimension in Flight. Technical report, Defense Advanced Research Projects Agency (DARPA). 1997

[6] Guerrero J. Wake signature and aerodynamic performance of finite-span root flapping rigid wings. Journal of Bionic Engineering. 2010;**7**:S109-S122

[7] Guerrero J, Pacioselli C, Pralits JO, Negrello F, Silvestri P, Lucifredi A, Bottaro A. Preliminary design of a small-sized flapping UAV I. Aerodynamic performance and static longitudinal stability. Meccanica. 2016;**51**:1343-1367

[8] Negrello F, Silvestri P, Lucifredi A, Guerrero J, Bottaro A. Preliminary design of a small-sized flapping UAV II. Kinematic and structural aspects. Meccanica. 2016;**51**:1369-1385

[9] Orchini A, Mazzino A, Guerrero J, Festa R, Boragno C. Flapping states of an elastically anchored plate in a uniform flow with applications to energy harvesting by fluid/structure interaction. Physics of Fluids. 2013;**25**:097105–1/097105–17

[10] Michelson R, Naqvi M. Extraterrestrail flight (entompter-based mars surveyor). Von Karman Institute for Fluid Dynamics. Lecture Series. November 2003;**2003**:24-28

[11] Triantafyllou GS, Triantafyllou MS. An efficient swimming machine. Scientific American. 1995;**272**:40-48

[12] Rohr J, Fish F. Strouhal number and optimization of swimming by odontocete cetaceans. The Journal of Experimental Biology. 2004;**207**:1633-1642

[13] Triantafyllou MS, Triantafyllou GS, Gopalkrishnan R. Wake mechanics for thrust generation in oscillating foils. Physics of Fluids. 1991;**3**:2835-2837

[14] Nudds RL, Taylor GK, Thomas AR. Tuning of strouhal number for high propulsive efficiency accurately predicts how wingbeat frequency and stroke amplitude relate and scale with size and flight speed in birds. Proceedings of the Biological Sciences. 2004;**7**:2071-2076

[15] Taylor GK, Nudds RL, Thomas AR. Flying and swimming animals cruise at a strouhal number tuned for high power efficiency. Letters to Nature. 2003;**425**:707-711

[16] Henshaw WD. A fourth-order accurate method for the incompressible Navier-Stokes equations on overlapping grids. Journal of Computational Physics. 1994;**113**:13-25

[17] Henshaw WD, Petersson NA. A split-step scheme for the incompressible Navier-Stokes equations. Numerical Solutions of Incompressible Flows. Preprint UCRL-JC-144. 2003. pp. 1-14

[18] Chesshire G, Henshaw W. Composite overlapping meshes for the solution of partial differential equations. Journal of Computational Physics. 1990;**90**:1-64

[19] Petersson N. Stability of pressure boundary conditions for stokes and navier-stokes equations. Journal of Computational Physics. 2001;**172**:40-70

[20] Sani RL, Shen J, Pironneau O, Gresho PM. Pressure boundary condition for the time-dependent incompressible navier-stokes equations. International Journal for Numerical Methods in Fluids. 2006;**50**:673-682

[21] Guerrero J. Numerical Simulation of the Unsteady Aerodynamics of Flapping Flight PhD thesis. Italy: University of Genoa, Deparment of Civil, Enviromental and Architectural Engineering; 2009. pp. 1-220

[22] Vinokur M. Conservation equations of gas-dynamics in curvilinear coordinate systems. Journal of Computational Physics. 1974;**14**:105-125

[23] Schiesser WE. The Numerical Method of Lines: Integration of Partial Differential Equations. San Diego, California: Academic Press; 1991. pp. 1-326

[24] Wesseling P. Principles of Computational Fluid Dynamics. Berlin, Heidelberg: Springer; 2001. pp. 1-644

[25] Roache P. Verification and Validation in Computational Science and Engineering. Albuquerque, New Mexico: Hermosa Publishers; 1998. pp. 1-464

[26] Roache P. Quantification of uncertainty in computational fluid dynamics. Annual Review of Fluid Mechanics. 1997;**29**:123-160

[27] Jeong J, Hussain F. On the identification of a vortex. Journal of Fluid Mechanics. 1995;**285**: 69-94

[28] Haimes R, Kenwright D. On the velocity gradient tensor and fluid feature extraction. In: AIAA Paper 1999–3288-CP. 1999. pp. 1-10

[29] Wang ZJ. Vortex shedding and frequency selection in flapping flight. Journal of Fluid Mechanics. 2000;**410**:323-341

[30] Young J, Lai J. Oscillation frequency and amplitude effects on the wake of a plunging airfoil. AIAA Journal. 2004;**42**:2042-2052

[31] Guerrero J. Aerodynamic performance of cambered heaving airfoils. AIAA Journal. 2010; **48**:2694-2698

[32] Parker K, Soria J, von Ellenrieder K. Thrust measurements from a finite-span flapping wing. AIAA Journal. 2007;**45**:58-70

[33] Parker K, von K, Soria J. Flow structures behind a heaving and pitching finite-span wing. Journal of Fluid Mechanics. 2003;**490**:129-138

[34] Dong H, Mittal R, Najjar FM. Wake topology and hydrodynamic performance of low-aspect-ratio flapping foils. Journal of Fluid Mechanics. 2006;**566**:309-343

[35] Blondeaux P, Fornarelli F, Guglielmini L, Triantafyllou MS, Verzicco R. Numerical experiments on flapping foils mimicking fish-like locomotion. Physics of Fluids. 2005;**17**: 113601-113612

[36] Young J, Lai J. Mechanisms influencing the efficiency of oscillating airfoil propulsion. AIAA Journal. 2007;**45**:1695-1702

Chapter 8

Aeroelastic Stability of Turboprop Aircraft: Whirl Flutter

Jiří Čečrdle

Additional information is available at the end of the chapter

http://dx.doi.org/10.5772/intechopen.70171

Abstract

This chapter is focused on a specific type of dynamic aeroelastic stability phenomenon—whirl flutter. Whirl flutter is caused by the effect of rotating parts of a turboprop power plant (propeller and, gas turbine engine rotor). The chapter presents fundamental facts regarding the whirl flutter phenomenon, including a historical overview and information regarding the occurrence of whirl flutter in aerospace practice. After that, the physical principles of whirl flutter are explained using a simple mechanical system with two degrees of freedom. Next, an analytical solution to determine the aerodynamic forces caused by the gyroscopic motion on each of the propeller blades is provided and the influences of the main structural parameters on the whirl flutter stability are discussed. The second, practical part is focused on the experimental research of the whirl flutter phenomenon and on the certification-related issues. The methodology of certification according to the FAR/CS 23 regulation standard is demonstrated on the example of a twin wing mounted tractor engine commuter aircraft.

Keywords: aeroelasticity, flutter, whirl flutter, quasi-steady theory, gyroscopic system

1. Introduction

Whirl flutter instability is a specific type of aeroelastic flutter instability that may appear on turboprop aircraft. It accounts for the dynamic and aerodynamic effects of rotating parts, such as a gas turbine engine rotor or a propeller. The rotating mass increases the number of degrees of freedom and generates additional forces and moments. Rotating propellers also cause an aerodynamic interference effect between a propeller and the structure of a nacelle and a wing. Whirl flutter instability is driven by motion-induced unsteady aerodynamic propeller forces and moments acting on the propeller plane. It may cause unstable vibration, which can lead to failure of an engine installation or an entire wing.

The propeller whirl flutter phenomenon was analytically discovered by Taylor and Browne [1]. The next pioneering work was performed by Ribner, who set the basic formulae for the aerodynamic derivatives of propeller forces and moments due to the motion and velocities in pitch and yaw in 1945 [2, 3]. After the accidents of two Lockheed L-188 C Electra II airliners in 1959 and 1960 [4], the importance of the whirl flutter phenomenon in practical applications was recognised.

This chapter is focused on turboprop aircraft whirl flutter; however, it may also occur in tilt-rotor aircraft. The whirl flutter phenomenon relates the mutual interactions of the rotating propeller with the aircraft deformations and the aerodynamic forces emerging during forward flight.

2. Physical principle and analytical solution

The principle of the whirl flutter phenomenon is outlined on a simple mechanical system with two degrees of freedom [5]. The propeller and hub are considered to be rigid. A flexible engine mounting is substituted with a system of two rotational springs (of stiffnesses K_Ψ and K_Θ), as illustrated in **Figure 1**.

Such a system has two independent mode shapes of yaw and pitch, with respective angular frequencies of ω_Ψ and ω_Θ, as shown in **Figure 2**.

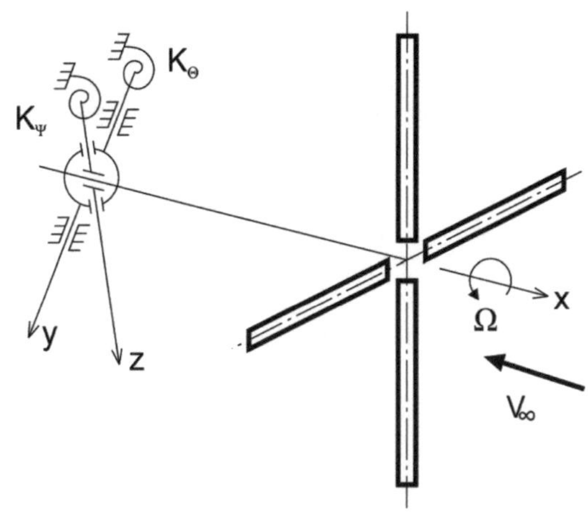

Figure 1. Two-degree-of-freedom gyroscopic system with a propeller.

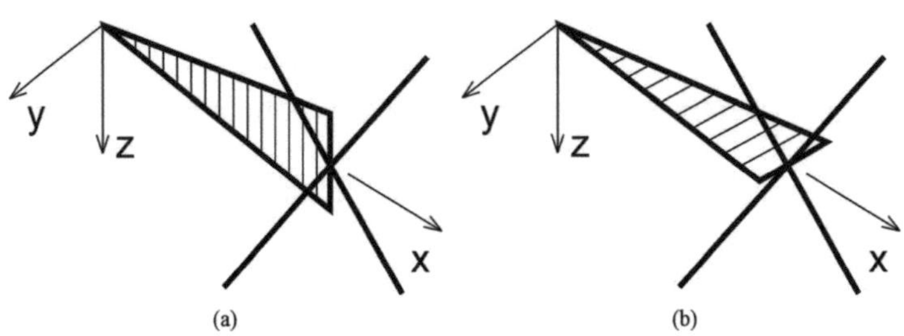

Figure 2. Independent engine pitch (a) and yaw (b) mode shapes.

Considering the propeller rotation with angular velocity Ω, the primary system motion changes to the characteristic gyroscopic motion. The gyroscopic effect causes two independent mode shapes to merge into whirl motion as shown in **Figure 3**. The axis of a propeller makes an elliptical movement. The orientation of the propeller axis movement is backward relative to the propeller rotation for the lower-frequency mode (backward whirl mode) and is forward relative to the propeller rotation for the higher-frequency mode (forward whirl mode). The trajectory of this elliptical movement depends on both angular frequencies ω_Ψ and ω_Θ. Because the yaw and pitch motions have a 90° phase shift, the mode shapes in the presence of gyroscopic effects are complex.

The described gyroscopic motion causes the angles of attack of the propeller blades to change, which consequently leads to unsteady aerodynamic forces. These forces may, under specific conditions, induce whirl flutter instability. The most important terms regarding whirl flutter are yaw moment due to pitch $M_Z(\Theta)$ and, similarly pitch moment due to yaw $M_Y(\Psi)$. These moments are to be balanced by aerodynamic or structural damping terms. The state of neutral stability with no damping of the system represents the flutter state. The corresponding airflow ($V_\infty = V_{FL}$) is called the critical flutter speed. In terms of flutter, the stable and unstable states of the gyroscopic system are applicable. Both states for the backward mode are explained in **Figure 4**. As long as the air velocity is lower than a critical value ($V_\infty < V_{FL}$), the system is stable

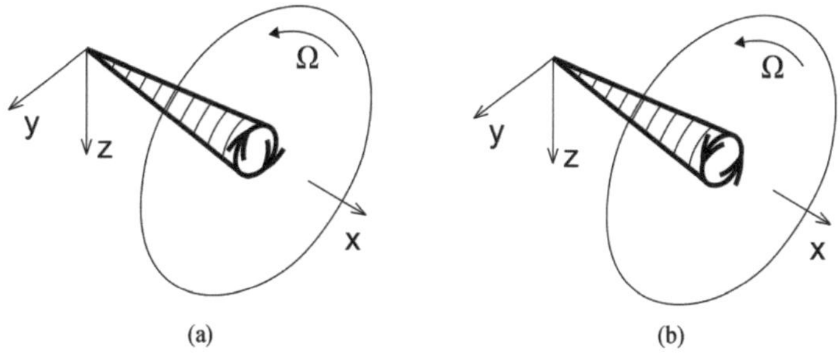

Figure 3. Backward (a) and forward (b) whirl modes.

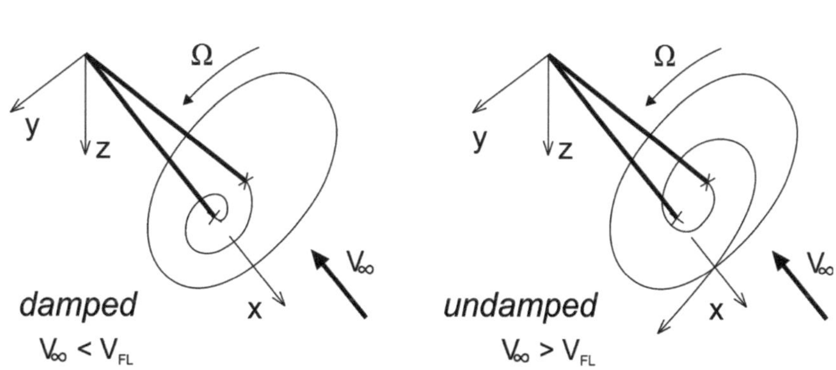

Figure 4. Stable (a) and unstable (b) states of gyroscopic vibrations for the backward mode.

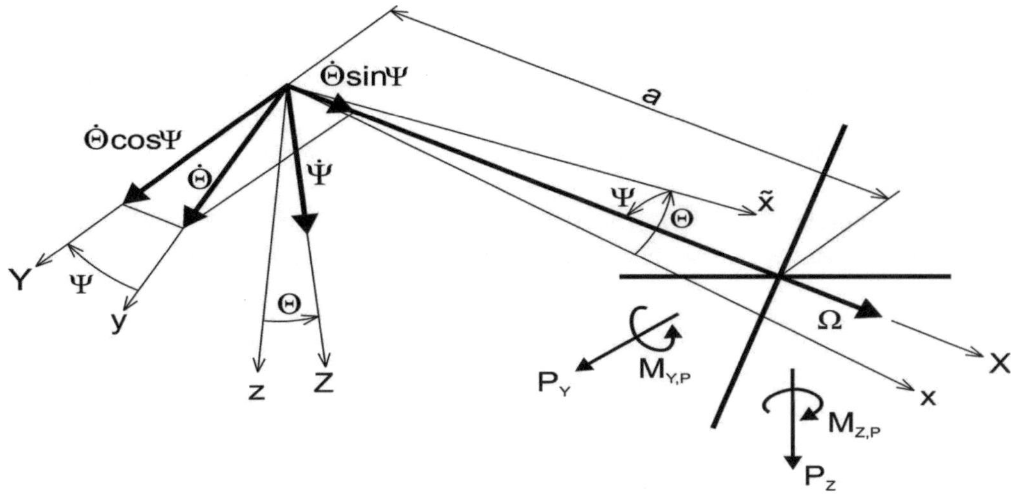

Figure 5. Kinematical scheme of the gyroscopic system.

and the gyroscopic motion is damped. When the airspeed exceeds the critical value ($V_\infty > V_{FL}$), the system becomes unstable and the gyroscopic motion is divergent.

The main problem in obtaining the analytical solution is to determine the aerodynamic force caused by the gyroscopic motion on each of the propeller blades. The presented equations of motion were set up for the system described in **Figure 1** by means of Lagrange's approach. The kinematical scheme, including gyroscopic effects, is shown in **Figure 5**.

Three angles (φ, Θ, Ψ) are selected as the independent generalised coordinates. The propeller is assumed to be cyclically symmetric with respect to both mass and aerodynamics (i.e., a propeller with a minimum of three blades). The angular velocity is considered to be constant ($\varphi = \Omega t$). Nonuniform mass moments of inertia of the engine with respect to the yaw and pitch axes ($J_Y \neq J_Z$) are also considered. We will use a coordinate system *X*, *Y*, *Z* linked to the system. Then, kinetic energy is:

$$E_K = \frac{1}{2} J_X \omega_X^2 + \frac{1}{2}\left(J_Y \omega_Y^2 + J_Z \omega_Z^2\right) \tag{1}$$

The angular velocities will be

$$\begin{aligned} \omega_X &= \Omega + \dot{\Theta}\sin\Psi \approx \Omega + \dot{\Theta}\Psi \\ \omega_Y &= \dot{\Theta}\cos\Psi \approx \dot{\Theta} \\ \omega_Z &= \dot{\Psi} \end{aligned} \tag{2}$$

Considering that $\dot{\Theta}^2 \dot{\Psi}^2 << \Omega^2$, the equation for the kinetic energy becomes

$$E_K = \frac{1}{2} J_X \Omega^2 + J_X \Omega \Psi \dot{\Theta} + \frac{1}{2}\left(J_Y \dot{\Theta}^2 + J_Z \dot{\Psi}^2\right) \tag{3}$$

The first part of Eq. (3) is independent of both Θ and Ψ; thus, it does not appear in Lagrange's equation. Then, the potential energy becomes

$$E_P = \frac{1}{2}K_\Theta\Theta^2 + \frac{1}{2}K_\Psi\Psi^2 \tag{4}$$

To describe the damping, we assume the structural damping commonly used in the flutter analyses, with the damping force proportional to the amplitude of the displacement:

$$D = \frac{1}{2}\frac{K_\Theta\gamma_\Theta}{\omega}\dot{\Theta}^2 + \frac{1}{2}\frac{K_\Psi\gamma_\Psi}{\omega}\dot{\Psi}^2 \tag{5}$$

Then, we obtain from Lagrange's equations and Eqs. (3)–(5) a system of two mutually influencing differential equations:

$$\begin{aligned} J_Y\ddot{\Theta} + \frac{K_\Theta\gamma_\Theta}{\omega}\dot{\Theta} + J_X\Omega\dot{\Psi} + K_\Theta\Theta = Q_\Theta \\ J_Z\ddot{\Psi} + \frac{K_\Psi\gamma_\Psi}{\omega}\dot{\Psi} - J_X\Omega\dot{\Theta} + K_\Psi\Psi = Q_\Psi \end{aligned} \tag{6}$$

Generalized propeller forces and moments (see **Figure 5**) can be expressed as

$$\begin{aligned} Q_\Theta &= M_{Y,P} - aP_Z \\ Q_\Psi &= M_{Z,P} + aP_Y \end{aligned} \tag{7}$$

The index P means that the moment around the specific axis is at the plane of the propeller rotation. Employing the quasi-steady theory, the effective angles become

$$\begin{aligned} \Theta^* &= \Theta - \frac{\dot{Z}}{V_\infty} = \Theta - \frac{a\dot{\Theta}}{V_\infty} \\ \Psi^* &= \Psi - \frac{\dot{Y}}{V_\infty} = \Psi - \frac{a\dot{\Psi}}{V_\infty} \end{aligned} \tag{8}$$

Neglecting the aerodynamic inertia terms ($\dot{\Theta}^* \approx \dot{\Theta}$, $\dot{\Psi}^* \approx \dot{\Psi}$), we obtain the equations for the propeller's dimensionless forces and moments as follows:

$$\begin{aligned} P_Y &= q_\infty F_P\left(c_{y\Psi}\Psi^* + c_{y\Theta}\Theta^* + c_{yq}\frac{\dot{\Theta}^* R}{V_\infty}\right) \\ P_Z &= q_\infty F_P\left(c_{z\Theta}\Theta^* + c_{z\Psi}\Psi^* + c_{zr}\frac{\dot{\Psi}^* R}{V_\infty}\right) \\ M_{Y,P} &= q_\infty F_P D_P\left(c_{m\Psi}\Psi^* + c_{mq}\frac{\dot{\Theta}^* R}{V_\infty}\right) \\ M_{Z,P} &= q_\infty F_P D_P\left(c_{n\Theta}\Theta^* + c_{nr}\frac{\dot{\Psi}^* R}{V_\infty}\right) \end{aligned} \tag{9}$$

where F_P is the propeller disc area and D_P is the propeller diameter.

The aerodynamic derivatives representing the derivatives of the two aerodynamic forces and two aerodynamic moments with respect to the pitch and yaw angles and to the pitch and yaw angular velocities are then defined as follows:

$$\begin{aligned}
c_{y\Theta} &= \frac{\partial c_y}{\partial \Theta^*} & c_{y\Psi} &= \frac{\partial c_y}{\partial \Psi^*} & c_{yq} &= \frac{\partial c_y}{\partial\left(\frac{\dot{\Theta}R}{V_\infty}\right)} & c_{yr} &= \frac{\partial c_y}{\partial\left(\frac{\dot{\Psi}R}{V_\infty}\right)} \\
c_{z\Theta} &= \frac{\partial c_z}{\partial \Theta^*} & c_{z\Psi} &= \frac{\partial c_z}{\partial \Psi^*} & c_{zq} &= \frac{\partial c_z}{\partial\left(\frac{\dot{\Theta}R}{V_\infty}\right)} & c_{zr} &= \frac{\partial c_z}{\partial\left(\frac{\dot{\Psi}R}{V_\infty}\right)} \\
c_{m\Theta} &= \frac{\partial c_m}{\partial \Theta^*} & c_{m\Psi} &= \frac{\partial c_m}{\partial \Psi^*} & c_{mq} &= \frac{\partial c_m}{\partial\left(\frac{\dot{\Theta}R}{V_\infty}\right)} & c_{mr} &= \frac{\partial c_m}{\partial\left(\frac{\dot{\Psi}R}{V_\infty}\right)} \\
c_{n\Theta} &= \frac{\partial c_n}{\partial \Theta^*} & c_{n\Psi} &= \frac{\partial c_n}{\partial \Psi^*} & c_{nq} &= \frac{\partial c_n}{\partial\left(\frac{\dot{\Theta}R}{V_\infty}\right)} & c_{nr} &= \frac{\partial c_n}{\partial\left(\frac{\dot{\Psi}R}{V_\infty}\right)}
\end{aligned} \tag{10}$$

These aerodynamic derivatives can be obtained analytically [2, 3, 6] or experimentally. Considering the symmetry, they can be expressed as follows:

$$\begin{aligned}
&c_{z\Psi} = c_{y\Theta};\ c_{m\Psi} = -c_{n\Theta};\ c_{mq} = c_{nr};\ c_{zr} = c_{yq}; \\
&c_{z\Theta} = -c_{y\Psi};\ c_{n\Psi} = c_{m\Theta};\ c_{mr} = -c_{nq};\ c_{yr} = -c_{zq}
\end{aligned} \tag{11}$$

Neglecting the low value derivatives, we can consider:

$$c_{mr} = -c_{nq} = 0;\ c_{yr} = -c_{zq} = 0 \tag{12}$$

By substituting Eq. (10) into the equations of motion (Eq. (6)) and considering the harmonic motion

$$[\Theta, \Psi] = [\overline{\Theta}, \overline{\Psi}]e^{j\omega t} \tag{13}$$

we obtain the final whirl flutter matrix equation

$$\left(-\omega^2[M] + j\omega\left([D] + [G] + q_\infty F_P \frac{D_P^2}{V_\infty}[D^A]\right) + \left([K] + q_\infty F_P D_P [K^A]\right)\right)\left|\begin{matrix}\overline{\Theta} \\ \overline{\Psi}\end{matrix}\right| = \{0\} \tag{14}$$

where the mass matrix becomes

$$[M] = \begin{bmatrix} J_Y & 0 \\ 0 & J_Z \end{bmatrix} \tag{15}$$

the structural damping matrix becomes

$$[D] = \begin{bmatrix} \frac{K_\Theta \gamma_\Theta}{\omega} & 0 \\ 0 & \frac{K_\Psi \gamma_\Psi}{\omega} \end{bmatrix} \tag{16}$$

the gyroscopic matrix becomes

$$[G] = \begin{bmatrix} 0 & J_X\Omega \\ -J_X\Omega & 0 \end{bmatrix} \tag{17}$$

the structural stiffness matrix becomes

$$[K] = \begin{bmatrix} K_\Theta & 0 \\ 0 & K_\Psi \end{bmatrix} \tag{18}$$

the aerodynamic damping matrix becomes

$$\left[D^A\right] = \begin{bmatrix} -\frac{1}{2}c_{mq} - \frac{a^2}{D_P^2}c_{z\Theta} & \frac{1}{2}\frac{a}{D_P}c_{yq} - \frac{a}{D_P}c_{n\Theta} - \frac{a^2}{D_P^2}c_{y\Theta} \\ -\frac{1}{2}\frac{a}{D_P}c_{yq} + \frac{a}{D_P}c_{n\Theta} + \frac{a^2}{D_P^2}c_{y\Theta} & -\frac{1}{2}c_{mq} - \frac{a^2}{D_P^2}c_{z\Theta} \end{bmatrix} \tag{19}$$

and the aerodynamic stiffness matrix becomes

$$\left[K^A\right] = \begin{bmatrix} \frac{a}{D_P}c_{z\Theta} & c_{n\Theta} + \frac{a}{D_P}c_{y\Theta} \\ -c_{n\Theta} - \frac{a}{D_P}c_{y\Theta} & \frac{a}{D_P}c_{z\Theta} \end{bmatrix} \tag{20}$$

Equation (14) can be solved as an eigenvalue problem. The critical state emerges for a specific combination of the parameters V_∞ and Ω, for which the angular velocity ω becomes real.

The influences of the main structural parameters are shown in the next figures. **Figure 6** shows the influence of the propeller advance ratio ($V_\infty/(\Omega R)$) on the stability of an undamped gyroscopic system. Increasing the propeller advance ratio has a destabilising effect. Another important parameter is the propeller hub distance ratio (a/R), the influence of which is documented in **Figure 7**. **Figure 7** also shows the influence of the structural damping (γ), which is a significant stabilisation factor. In contrast, the influence of the propeller thrust is negligible. The most critical state is $\omega_\Theta = \omega_\Psi$, when the interaction of both independent yaw and pitch motions is maximal and the trajectory of the gyroscopic motion is circular. Considering rigid propeller blades, the whirl flutter inherently appears in the backward gyroscopic mode. The flutter frequency is the same as the frequency of the backward gyroscopic mode. The critical state can be reached by increasing either V_∞ or Ω. A special case of eq. (14) for $\omega = 0$ is gyroscopic static divergence, which is characterised by unidirectional divergent motion.

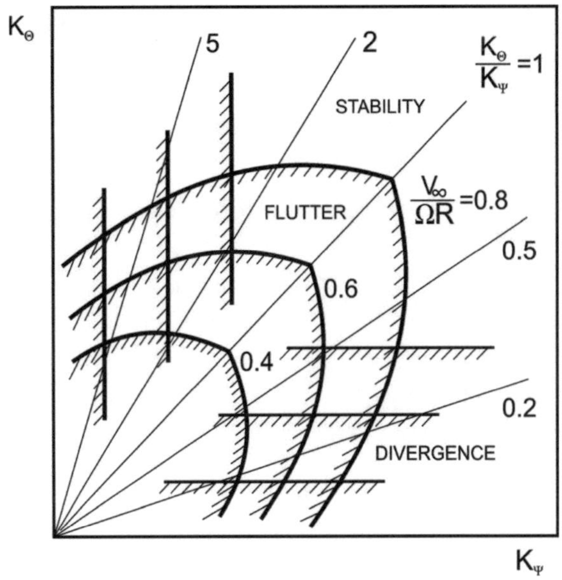

Figure 6. Influence of the propeller advance ratio on the stability of an undamped gyroscopic system.

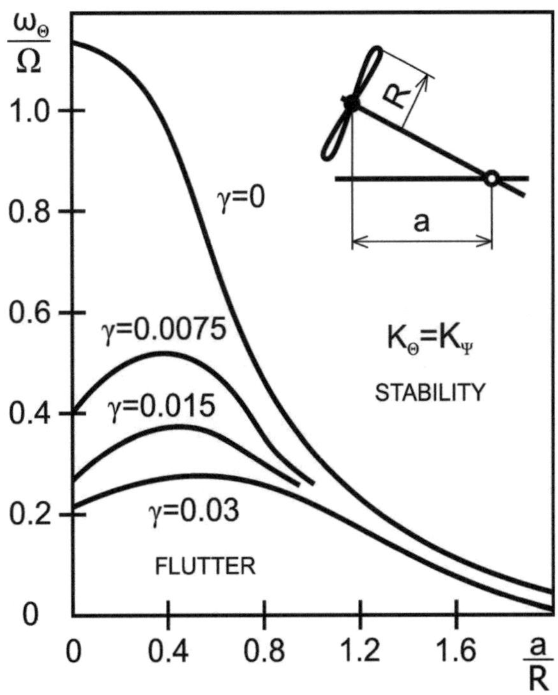

Figure 7. Influence of the propeller hub distance ratio.

http://dx.doi.org/10.5772/intechopen.70171

The described model, which is based on the assumption of a rigid propeller, is obviously applicable for standard turboprop aircraft (commuters, utility aircraft, and military trainers), for which the natural frequencies of the propeller blades are much higher than the frequencies of the engine system suspension vibrations. In large turboprops, in particular military transport aircraft with heavy multiblade propellers, the solution requires taking into account the deformations of the propeller blades as well [7–10]. Obviously, whirl flutter investigation of tilt-rotor aircraft must include even more complex analytical models [11, 12].

3. Experimental research

The analytically obtained results on the complicated physical principles of whirl flutter require experimental validation of the analytically obtained results, especially due to the unreliable analytical solution of the propeller aerodynamic forces. In addition, structural damping is a key parameter, to which whirl flutter is extremely sensitive and the characteristics of which need to be validated. Therefore, experiments using aeroelastic models are required.

Most of the developments in the whirl flutter experimental research were accomplished in the early 1960s. The experiments were carried out in direct connection with the Electras' accidents. The first experimental investigations were accomplished by Houbolt and Reed [6]. They used the simple model of a propeller in the windmilling mode. The complex investigations of whirl flutter characteristics were conducted by Bland and Bennett [13]. The measurements, which were carried out in the NASA Langley wind tunnel were focused on the propeller forces and stability of the propeller-nacelle component model. As was typical, a propeller rotated in the windmilling mode. The experimental results showed that the theoretical aerodynamic derivatives underestimated the whirl flutter speed and the application of the experimentally obtained derivatives to the analytical solution made the solution much closer to the experimental results. The tilt-rotor concept was researched by Reed and Bennett [14], who focused on the flight regimes of high inflow angles. Apart from the rigid blades, they also accounted for blades' flexibility when conducting the experiment on a simple model with flapping blades. Both backward and forward whirl mode instabilities occurred during the tests of the flapping-blade model. Similar experiments on the flapped blade rotor system were accomplished by Krishna Rao and Sundararajan [7] in NAL Bangalore. The influence of the blades' flapping frequency on the whirl flutter stability for both backward and forward whirl modes was demonstrated by these experiments.

A more complex model, which represented an aircraft half-wing with an engine, was tested by Bland and Bennett [15] in NASA Langley. That model was a typical aeroelastic model with a duralumin spar and balsa segment structure. The main focus of these experiments was to investigate the influence of wing stiffness on whirl flutter.

The largest experimental campaign was accomplished as a part of the response to the L-188 C Electra II aircraft accidents. Eventually, the tests helped to determine the cause of the accidents: whirl flutter. The aeroelastic model included four nacelles and four windmilling propellers. The model represented a full-span aircraft due to the investigation of the unsymmetrical phenomena. The model was flown in a wind tunnel with sufficient lift during the measurements and the trimmed flight was maintained by an operator controlling the horizontal stabilizer. Various configurations with reduced starboard outboard engine attachment stiffness were tested with the aim of identifying the causes of the aircraft accidents. The reduced stiffness parameters were also tested on the inboard power plant, and the combination of two engines was also tested. States with reduced damping were also tested. The experiments are summarised by Abbott et al. [16].

Further experimental activities were primarily focused on the issues connected with the design and development of tilt-rotor aircraft, such as the Bell XV-3 or XV-15, Bell-Boeing V-22 Osprey, Agusta-Westland AW609, and XC-142A. A large investigation of the proprotor research model with flapping blades was conducted by Kvaternik and Kohn [17]. The stiffness of the nacelle attachment was reduced to reach the flutter boundaries within low velocities. The main aim of the work was the necessity to establish an experimental database for proprotor whirl flutter prediction with sufficient confidence. A total of 26 backward whirl flutter states and 50 forward whirl flutter states were found.

Recent experimental studies include the work performed by Rand and Peyran [18]. The tests were aimed at assessing the effects of the noted structural characteristics couplings on the whirl flutter of the proprotor during forward flight. In addition, the possibility of suppressing the instability by means of the active control of the wing's structural characteristics was tested. The demonstrator included a proprotor in windmilling mode attached to the wing structure. A very simple table-top model was used by Acree et al. [19]. The model included weights in front of the leading edge of the blades' tips. Moving the weights chordwise caused significant changes in the whirl mode stability.

A large model of the tilt-rotor aircraft concept, including a half-wing and proprotor, was tested for whirl flutter in the DNW wind tunnel in the Netherlands. The model was based on an aerodynamic wind tunnel model of a previous project. Therefore, aeroelastic scaling and modifications were limited. A multibody analysis based on available technical information was reported by Krueger [20].

Experimental activities that employed a model of a tilt-rotor aircraft component or a complete tilt-rotor aircraft model were performed on the WRATS aeroelastic demonstrator. The complete model was based on the 1/5-size semispan aeroelastic model of the V-22 Osprey aircraft. The model was used during the development of the Osprey to improve the stability characteristics of the aircraft. Later, the model was modified and utilized as the research demonstrator for active control research. WRATS-related activities are summarised by Piatak and Kvaternik et al. [21] and by Nixon et al. [22].

The latest experimental activities were accomplished by Cecrdle et al. using the W-WING aeroelastic demonstrator [23]. The demonstrator was adapted from a half-wing with a span of 2.56 m with the engine of a former aeroelastic model of a commuter aircraft for 40 passengers. The total mass of the model is approximately 55.5 kg. The stiffness of the wing and aileron is modeled by a duralumin spar of variable cross-section. The aerodynamic shape is covered with modular balsa and plastic segments. The inertia characteristics are modeled by lead weights.

The aileron actuation stiffness is modeled by means of a replaceable steel spiral spring. Optionally, the aileron may be actuated by the hydraulic actuator or by the electromagnetic shaker placed at the wing root via a push-pull rod. Furthermore, an active control system that is capable of simulating the additional mass, damping, or stiffness, including the nonlinear characteristics of these terms [24], may also be applied. The wing is fixed at the root to the pylon that is attached to the wind tunnel manipulator.

The nacelle structure may be used either separately or attached to the wing structure as described. The influence of changes in the main parameters on the whirl flutter may be simulated by the demonstrator. The nacelle model includes two degrees of freedom (engine yaw and pitch). The engine attachment stiffness parameters are modeled using cross spring pivots. The leaf springs are changeable. The stiffness parameters can be adjusted independently by replacing the spring leaves. Both pivots are independently movable in the direction of the propeller axis. This allows for the adjustment of the pivot points of both vibration modes, while the overall length of the nacelle and the propeller position remain the same. The engine inertia parameters are modeled by a replaceable and movable weight. The weight is used to preserve position of the center of gravity in case the pivot stations change. Optionally, it also enables the position of the center of gravity to be changed. Provided the nacelle is attached to the wing, the wing dynamic characteristics can also be adjusted to evaluate the influence of the wing structure on the whirl flutter. The design solution of the engine attachment is shown in **Figure 8**.

Sensor instrumentation of the wing includes strain gauges in the root and half-span sections that are configured to measure the torsional, vertical bending, and in-plane bending deformations. In addition, the demonstrator is equipped with accelerometers at the front of the engine and at the wing-tip section. Accelerometers measure the vertical and lateral acceleration.

Figure 8. Design solution of W-WING demonstrator engine attachment, motor, and propeller.

The gyroscopic effect is simulated by the rotating mass of the propeller blades. Actually, two sets of blades are available (light, made of duralumin, and heavy, made of steel). The propeller of 0.7 m diameter represents a scaled-down real 5-blade propeller. The propeller is powered by an electric motor and it can operate at arbitrary revolutions of up to 3000 rpm. Obviously, the windmilling mode is also applicable. The propeller blades are adjustable at a standstill. There are several blade adjustment options (angles of attack) that are applicable for specific ranges of the flow velocity. Additionally, the blade angle of attack may be used to manage the revolutions of the windmilling propeller.

The tests were performed in the VZLU 3-m-diameter low-speed wind tunnel. To prevent the induced effects at the wing root region, the wing is combined with the splitter plate. The demonstrator is fixed to the attachment arm inside the wind tunnel test section. Both the angle of attack and angle of sideslip of the tested model may be changed, provided if requested. The test arrangement is shown in **Figure 9**.

The measurement variants of the model were defined by the following structural parameters: pitch and yaw attachment stiffness, pitch and yaw hinge station, mass-balance weight station, choice of propeller (duralumin or steel blades), and finally, the propeller blade's 75%-section angle of attack (α).

The tests were focused on the variation of the pitch and yaw stiffness first. During these tests, the most promising variants with respect to the pitch and yaw stiffness, which showed the largest vibrations, were found and used as the baselines during the next phase of the tests. Then, the variations of the blade angle of attack, choice of a light or heavy propeller, and variation of the pitch hinge station and mass-balance weight station were examined. The measurements included excitation by the flow turbulence and by the aileron flapping sweep. The latter was found as very useful for the estimation of whirl mode damping.

The next figures show examples of the experimental results. The figures show evaluated whirl mode parameters as functions of the windflow speed. These parameters are propeller revolutions,

Figure 9. W-WING demonstrator in the wind tunnel test section.

damping, and frequency of the whirl mode and the maximal amplitude (pitch or yaw) of the front engine sensor section. The first example (**Figure 10**) demonstrates a very stable case. The vibration amplitude of the structure is very low and the damping increases with the windflow velocity.

The next example (**Figure 11**) shows the case in which the instability was reached. The amplitude curve shows a rapid increase near the flutter velocity. Additionally, the damping reaches

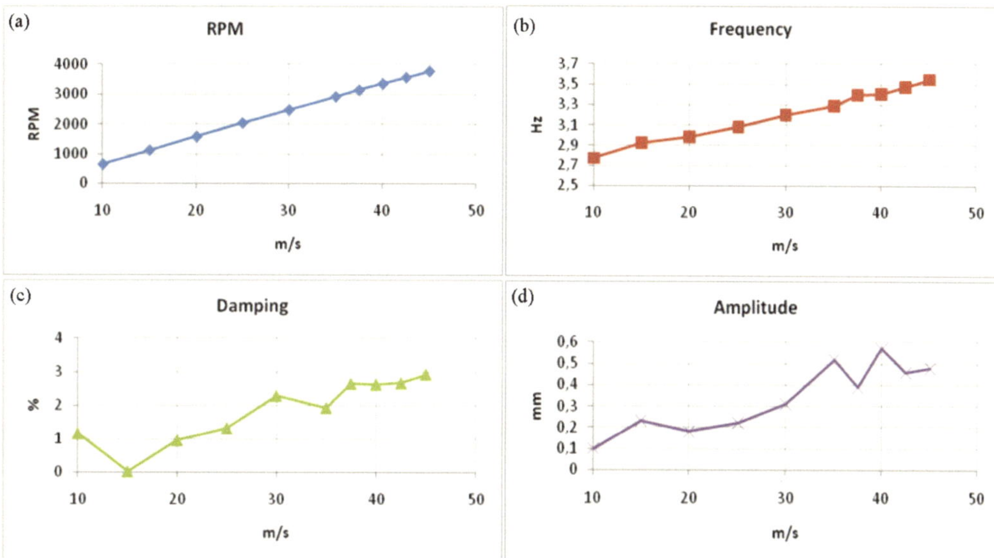

Figure 10. Whirl mode parameters: (a) propeller rpm, (b) whirl mode frequency, (c) whirl mode damping, (d) yaw or pitch amplitude. Stable case (pitch spring: nr.2, yaw spring: nr.2, pitch hinge: middle, weight: rear, blades: heavy, $\alpha = 2.5°$).

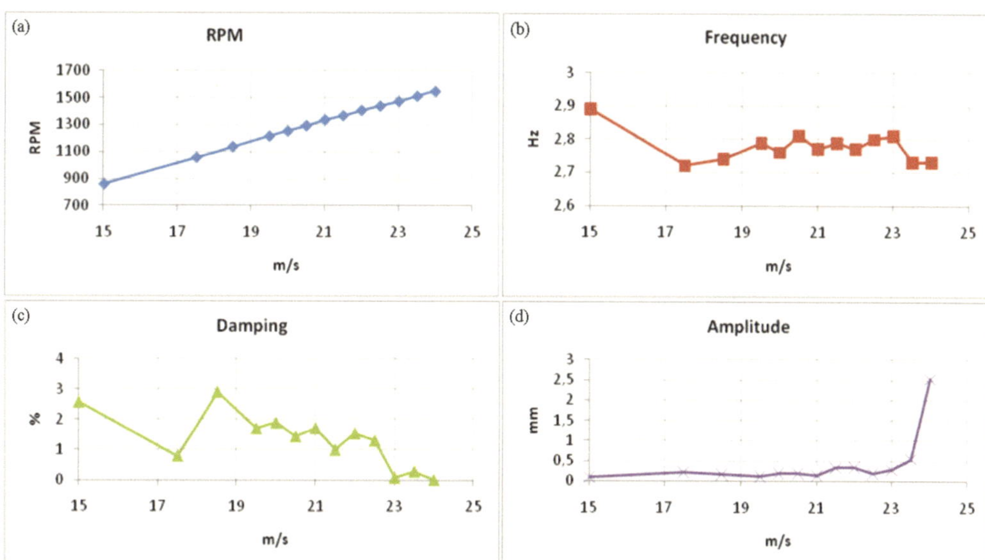

Figure 11. Whirl mode parameters: (a) propeller rpm, (b) whirl mode frequency, (c) whirl mode damping, (d) yaw or pitch amplitude. Unstable case (pitch spring: nr.2, yaw spring: nr.2, pitch hinge: middle, weight: front, blades: light, $\alpha = 0°$).

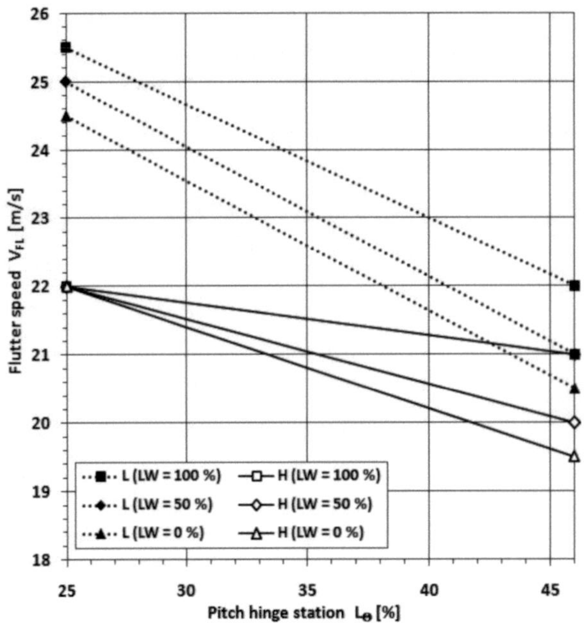

Figure 12. Influence of pitch hinge station (L_Θ) on flutter speed (25% = rear; 46% = middle; L = light blades; H = heavy blades; LW = mass-balance weight station).

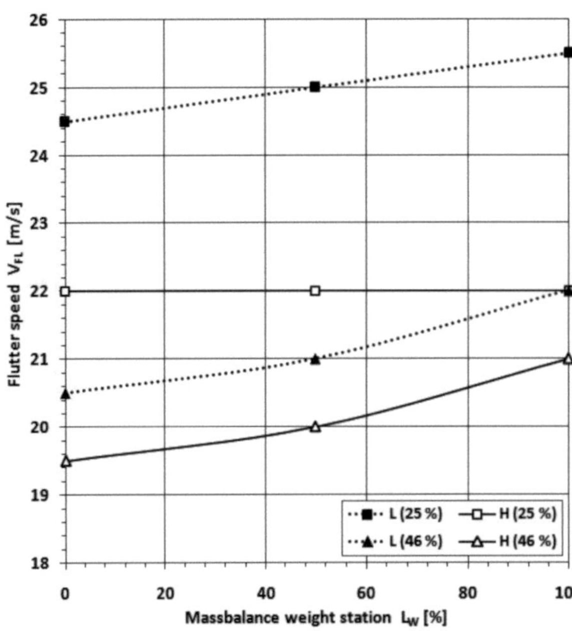

Figure 13. Influence of mass balance weight station (L_W) on flutter speed (0% = front; 100% = rear; L = light blades; H = heavy blades), rear (25%), and middle (46%) pitch hinge stations.

zero. The damping curve represents the damping values given by operational modal analysis (OMA). These damping values, which were obtained by the evaluation of the logarithmic decrement, are very small negative values, which are not noticeable in the figure. A negative damping value represents an unstable state. The unstable states with higher negative damping could not be reached due to safety reasons.

Figures 12 and **13** show more detailed information regarding the influence of the parameters. The influence of the pitch hinge station is demonstrated in **Figure 12**. Moving the hinge rearward increases the a/R ratio, and therefore, has a positive effect on flutter stability. The influence of the mass-balance weight station is demonstrated in **Figure 13**. Moving the weight rearward causes a decrease in mass moments of inertia (J_Y and J_Z) and the related increase in the yaw and pitch frequency (f_Ψ and f_Θ), and therefore, it also has a positive effect on flutter stability. Flutter speeds of configurations with heavy blades are lower compared to the flutter speeds with light blades as the heavy blades reduce both the yaw and pitch frequencies (f_Ψ and f_Θ).

4. Certification process

The airworthiness regulation standard requirements for dealing with the aeroelasticity and flutter also include requirements related to the whirl flutter. Essentially, the whirl flutter requirements are applicable only to aircraft powered by a turboprop power plant system. In the following text, the requirements of the FAR/CS 23 regulation standard, which is applicable to smaller turboprop aircraft, are taken into consideration. The whirl flutter related requirement, which is included in §629(e), is applicable regardless of the aircraft configuration or the number of engines (twin wing-mounted, single nose-mounted, twin fuselage mounted pusher, etc.). §629(e)(1) includes the main requirement to evidence the stability within the required V-H envelope, while §629(e)(2) requires the variation of structural parameters such as the stiffness and damping of the power plant attachment. The whirl flutter analysis must, therefore, include all significant aircraft configurations with respect to fuel and payload that are applicable to the aircraft operation. The whirl flutter analysis must also include the influence of the variance of the power plant mount structural parameters when simulating the possible changes due to structural damage (e.g., deterioration of engine mount isolators). Note that further requirements that are applicable to aircraft compliance with the fail-safe criteria come from §629(g).

We can use two main approaches for the analysis:

First is the standard approach, in which the analyses are performed for a known set of structural parameters, and the results are whirl flutter stability characteristics (e.g., whirl flutter speed). The resulting flutter speed is then compared to the certification velocity according to the flight envelope. The analyses are performed sequentially, state by state.

The standard approach is good for complying with the main requirement (§629(e)(1)), which is realized by evaluation of the nominal parameter states. **Figure 14** shows an example of a V-g-f diagram of such a calculation. No flutter instability is indicated up to the certification velocity, which is 191.4 m/s in the case, and therefore, the regulation requirement is fulfilled. Calculations are performed for all applicable mass configurations. However, for parametric studies that

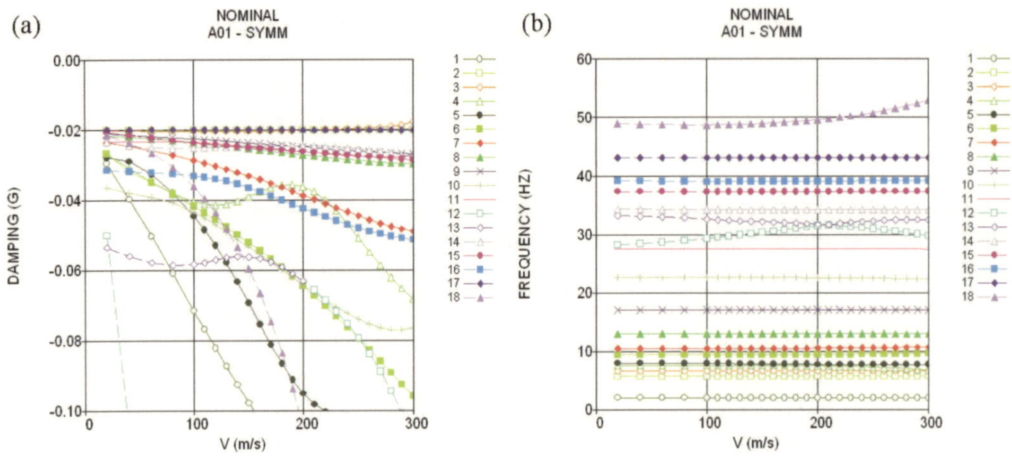

Figure 14. Example of whirl flutter calculation (V-g-f) diagram, (a) damping, (b) frequency, nominal state.

may include huge numbers of analyses, such an approach may become ineffective unless some tool for automated analysis, data handling, and processing is used. However, the applicability of such automatic processing systems is always limited.

Therefore, to comply with the parameter variation requirement (§629(e)(2)), the second, optimization-based approach [25] can be used. In this approach, the flutter speed is set equal to the certification speed, and the results are critical values of the structural parameters. The stability margin can then be obtained from these critical structural parameters. The analyzed states are then compared only with respect to the structural parameters and the relationship to the stability margin. Such an approach can save large amounts of time because the number of required whirl flutter analyses is dramatically reduced.

Provided a full-span model is considered, four design variables are defined: (1) effective stiffness of the engine attachment for symmetric pitch, (2) effective stiffness of the engine attachment for antisymmetric pitch, (3) effective stiffness of the engine attachment for symmetric yaw, and (4) effective stiffness of the engine attachment for antisymmetric yaw. The solution includes three frequency ratio constraints: (1) for symmetric engine vibration frequencies, (2) for antisymmetric engine vibration frequencies, and (3) for critical whirl flutter frequencies. Additionally, the flutter constraint, i.e., the requirement of flutter stability, is applied for the certification speed. The objective function is then formally expressed as the minimization of the sum of engine vibration frequencies. **Figure 15** shows an example of a V-g-f diagram for the optimization-based calculation. There is a flutter state of mode nr.2 (engine pitch vibration mode) at the velocity of 191.4 m/s representing the whirl flutter instability. Calculations are performed for several values of the critical frequency ratio to construct a stability margin curve, which is then constructed for all applicable mass configurations, as shown in the example in **Figure 16.** Stability margins may be constructed with respect to either engine yaw and pitch vibration frequency or engine yaw and pitch attachment effective stiffness. The former type of margin is then compared with the engine vibration frequencies, obtained by the GVT or analytically, to evaluate the rate of reserve. **Figure 17** demonstrates an example of such an evaluation. The

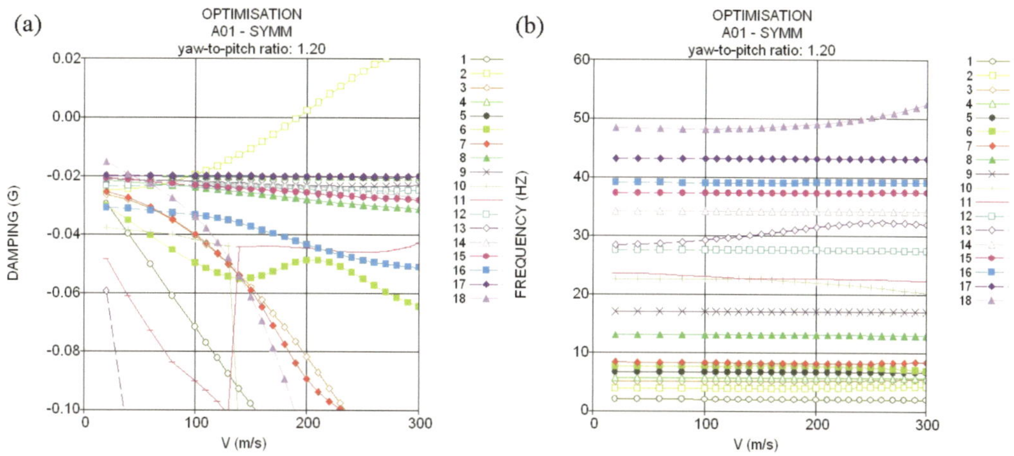

Figure 15. Example of whirl flutter calculation (V-g-f) diagram, (a) damping, (b) frequency, optimization-based calculation.

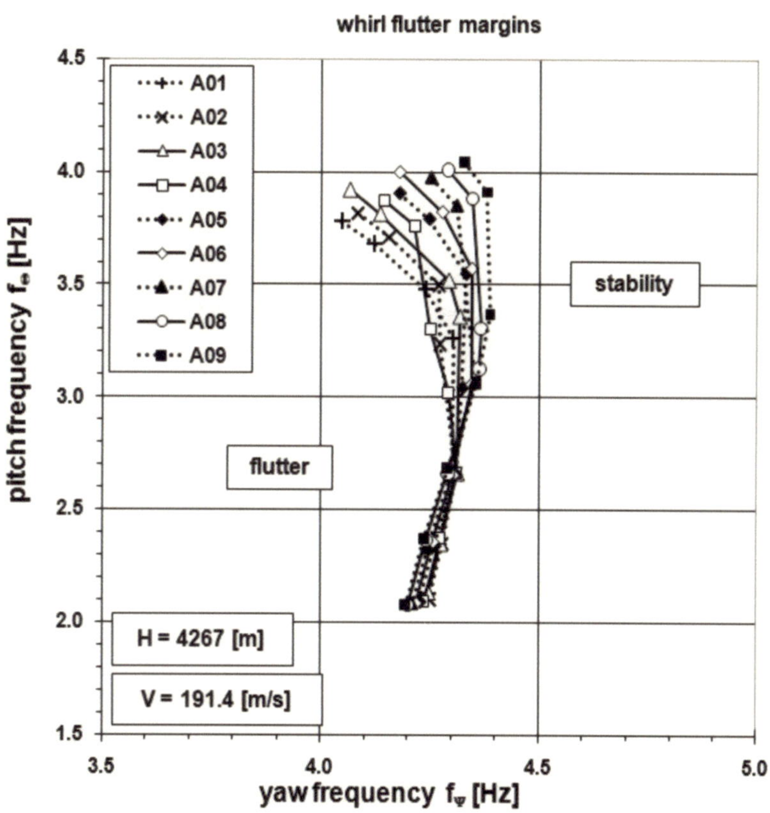

Figure 16. Example of whirl flutter stability margins for multiple mass configurations.

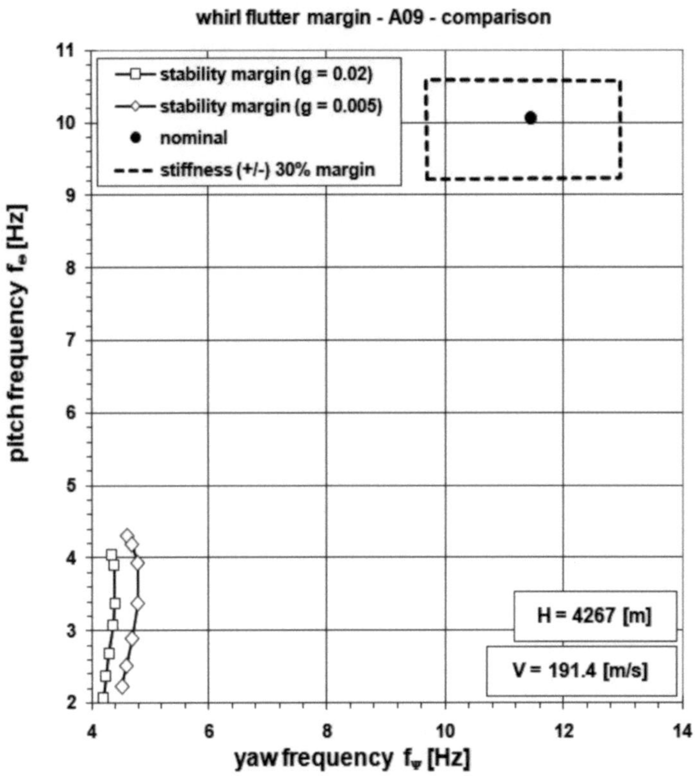

Figure 17. Example of whirl flutter stability margin evaluation.

dashed line represents the (±) 30% variance margin in engine attachment stiffness. Another parameter to be evaluated is the damping. This is provided by calculation with very low structural damping, represented by the damping of $g = 0.005$, while the standard structural damping included in the analyses is $g = 0.02$. The stability margin for reduced damping is also presented in **Figure 17**.

As obvious from **Figure 17**, there is sufficient reserve in stability of the nominal state with respect to the stability margin, and therefore, the regulation requirements would are fulfilled.

5. Conclusion

The presented chapter addresses a specific aeroelastic phenomenon that is applicable for turboprop aircraft structures: whirl flutter. This chapter includes basic facts regarding the physical principles and the analytical solution of the described phenomenon. After that, the experimental research activities are outlined, with a focus on the recent experiments on the W-WING whirl flutter demonstrator. Finally, the approaches to aircraft certification are explained. Comprehensive information on the whirl flutter phenomenon from all aspects can be found in Ref. [26].

Acknowledgements

This work was supported by the institutional support of nonprofit research organisations by the Ministry of Industry and Trade of the Czech Republic.

Author details

Jiří Čečrdle

Address all correspondence to: cecrdle@vzlu.cz

Aeronautical Research and Test Institute (VZLU), Prague, Czech Republic

References

[1] Taylor ES, Browne KA. Vibration isolation of aircraft power plants. Journal of Aerospace Sciences. 1938;**6**:43-49

[2] Ribner HS. Propellers in Yaw. Langley: NASA Report 820; 1945

[3] Ribner HS. Formulas for Propellers in Yaw and Charts of the Side-Force Derivatives. Langley: NASA Report 819; 1945

[4] Donham RE, Watts GA. Whirl flutter first case. In: Flomenhoft H, editor. The Revolution in Structural Dynamics. 1st ed. Palm Beach Gardens: Dynaflo Press; 1997. pp. 99-109

[5] Reed WH. Review of Propeller-Rotor Whirl Flutter. Langley: NASA Report TR R-264; 1967

[6] Houbolt JC, Reed WH. Propeller-Nacelle whirl flutter. Journal of Aerospace Sciences. 1962;**3**:333-346

[7] Krishna Rao KV, Sundararajan D. Whirl Flutter of Flapping Blade Rotor Systems. Bangalore: National Aeronautical Laboratory TN-18; 1969

[8] Richardson JR, Naylor HFW. Whirl Flutter of Propellers With Hinged Blades. Toronto: Engineering Research Associates Report No. 24; 1962

[9] Richardson JR, McKillop JA, Naylor HFW, Bandler PA. Whirl Flutter of Propellers with Flexible Twisted Blades. Toronto: Engineering Research Associates Report No. 43; 1962

[10] Donham RE. Effect of propeller blade bending dynamics on 1P loads and whirl flutter. In: International Forum on Aeroelasticity and Structural Dynamics; 28 June-1 July 2005; Munich

[11] Barkai SM, Rand O, Peyran RJ, Carlson RM. Modeling and analysis of tilt-rotor aeromechanical phenomena. Mathematical and Computer Modelling. 1998;**12**:17-43

[12] Dugeai A, Mauffrey Y, Sicot F. Aeroelastic capabilities of the elsA solver for rotating machines applications. In: International Forum on Aeroelasticity and Structural Dynamics; 26-30 June 2011; Paris

[13] Bland SR, Bennett RM. Wind-Tunnel Measurement of Propeller Whirl-Flutter Speeds and Static Stability Derivatives and Comparison with Theory. Langley: NASA Report TN D-1807; 1963

[14] Reed WH, Bennett RM. Propeller Whirl Flutter Considerations for V/STOL Aircraft. In: CAL/TRECOM Symposium; Buffalo; 1963

[15] Bennett RM, Bland SR. Experimental and Analytical Investigation of Propeller Whirl Flutter of a Power Plant on a Flexible Wing. Langley: NASA Report TN D-2399; 1964

[16] Abbott FT, Kelly HN, Hampton KD. Investigation of Propeller Power Plant Autoprecession Boundaries for a Dynamic-Aeroelastic Model of a Four-Engine Turboprop Transport Airplane. Langley: NASA Report TN D-1806; 1963

[17] Kvaternik RG, Kohn JS. An Experimental and Analytical Investigation of Proprotor Whirl Flutter. Langley: NASA Report TP 1047; 1977

[18] Rand O, Peyran RJ. Experimental demonstration of the influence of wing couplings on whirl–flutter instability. Journal of Aircraft. 2000;**5**:859-864

[19] Acree CW, Peyran RJ, Johnson W. Rotor design for whirl flutter: An examination of options for improving tiltrotor aeroelastic stability margins. In: American Helicopter Society 55th Annual Forum; 25-27 May 1999; Montreal

[20] Krueger WR. Analysis of whirl flutter dynamics on a tiltrotor wind tunnel model. In: International Forum of Aeroelasticity and Structural Dynamics; 21-25 June 2009; Seattle

[21] Piatak DJ, Kvaternik RG, Nixon MW, Langston ChW, Singleton JD, Bennett RL, Brown RK. A wind-tunnel parametric investigation of tiltrotor whirl-flutter stability boundaries. In: American Helicopter Society 57th Annual Forum; 9-11 May 2001; Washington

[22] Nixon MW, Langston ChW, Singleton JD, Piatak DJ, Kvaternik RG, Corso LM, Brown RK. Aeroelastic stability of a four-bladed semi articulated soft inplane tiltrotor model. In: American Helicopter Society 59th Forum; 6-8 May 2003; Phoenix

[23] Cecrdle J, Malinek P, Vich O. Wind tunnel test of whirl flutter aeroelastic demonstrator. In: 58th AIAA/ASCE/AHS/ASC Structures, Structural Dynamics, and Materials Conference; 9-13 January 2017; Grapevine. DOI: 10.2514/6.2017-0635

[24] Malecek J, Cecrdle J, Hlavaty V, Malinek P. Mechanical concepts for simulation of nonlinearities on aeroelastic demonstrator. Journal of Aircraft. 2013;**2**:651-658

[25] Cecrdle J. Analysis of twin turboprop aircraft whirl-flutter stability boundaries. Journal of Aircraft. 2012;**6**:1718-1725. DOI: 10.2514/1.C031390

[26] Cecrdle J. Whirl Flutter of Turboprop Aircraft Structures. 1st ed. Oxford: Elsevier Science; 2015. p. 354

Chapter 9

Goal- and Object-Oriented Models of the Aerodynamic Coefficients

Jozsef Rohacs

Additional information is available at the end of the chapter

http://dx.doi.org/10.5772/intechopen.71419

Abstract

Nowadays, aeronautics discovers new ways of flights near the critical regimes, unconventional aircraft forms, utilizing the micro–electro-mechanical technologies in flow and aircraft control, adaptive and morphing structures, using the structures and controls based on the biological principles, developing highly flexible structures, etc. Before deployment, these new technologies and solutions must be evaluated, tested in wide aerodynamic, flight dynamic simulations that require improved and new type of aerodynamic coefficient models. The chapter overviews the applicable models of the aerodynamic coefficients, introduces some new models and demonstrates how the different models can be applied in different goal- and object-oriented solutions. The following will be shortly explained: (i) how the aerodynamic forces and moments are generating, (ii) how the linear, nonlinear, steady, and nonsteady aerodynamic coefficient structures and forms might be modeled, and (iii) how to harmonize the model with the goal and object of investigations.

Keywords: aerodynamic coefficients, models of aerodynamic coefficients, critical regimes

1. Introduction

Aerodynamics is a fundamental subject investigating the interaction of the (atmospheric) gases with objects moving in them. This is a basic science that explains how to develop flying objects (aircraft) with minimum drag, maximum lift, and acceptable and controllable side force and moments.

Aerodynamics [1–8] deals with the theory of aerodynamic force and moment generation and the description of force and moment components appearing on aerofoils, wings, rotating wings, circular bodies at low, moderate, subsonic, supersonic, and hypersonic speeds and developing the models and methods of calculating the aerodynamic forces and moments. The theoretical and practical methods of evaluation and estimation of the aerodynamic forces and moments are synthetized in aircraft aerodynamic design, i.e., finding the best aerodynamic shape of the aircraft with maximum lift and minimum drag (ratio of which is called as aerodynamic goodness) and controllable other force moments. The aerodynamic characteristics are applied in aircraft motion description, namely for estimating the flight performance, determining the stability conditions and stability, flight dynamics and control.

Aerodynamics is a subfield of fluid and gas dynamics and uses their basic equations. However, there are no good and general methods for calculating the aerodynamic forces and moments that depend on shape and geometrical characteristics of the body, fluid properties, and motion dynamics. Therefore, a series of nondimensional aerodynamic coefficients were introduced, and with the use of results from theoretical and practical investigations (including the computation fluid dynamics and wind tunnel and flight tests), different models of aerodynamic coefficients were developed. The models depend on the real situations, objects, and goals of their application as shown in **Figure 1**, reflecting the aerodynamic mathematical modeling approach of Tobak [11] in the form of known Bisplinghoff's representation [12].

This chapter describes the goal- and object-oriented models of the aerodynamic coefficients and discusses their applicability. It contains 10 subchapters (10 points). The first is this introduction. The second one shortly explains the aerodynamic force and moment generation. The third point introduces the aerodynamic coefficients and defines their mathematical models. The fourth subchapter deals with the first, simple models based on several partial derivatives. The fifth point states improvement of the simple models and describes the so-called classic aerodynamic models. Generally, these models are most used by aerodynamics, flight performance, stability, flight dynamics, and control. The developed aerodynamic models described

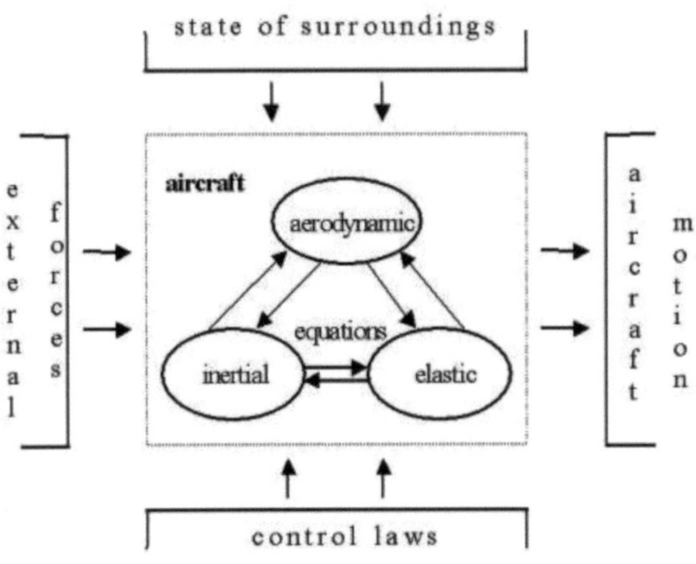

Figure 1. Modeling approach to aerodynamic coefficients (affecting aspects) [9, 10].

http://dx.doi.org/10.5772/intechopen.71419

by the next point are used for nonsteady aerodynamics, studying the very nonlinear and even chaotic motion of aircraft. The seventh point shows how the advanced aerodynamic coefficient might be created including the analytical models, special approximation of the measured data, using the soft computing to estimate the coefficient models. Applicability of the described models is discussed in the following eighth subchapter. Finally, the ninth point shows a use of an advanced aerodynamic model. The conclusion (tenth point) summarizes the materials introduced and described by this chapter.

2. Aerodynamic force and moment generation

Aerodynamic force and moment are represented and investigated by their components due to the applied reference (coordinate) system. There are several reference systems used. When investigating the stability and control [13–15], the usual body reference is applied, where the center of a right hand Descartes system is located at the aircraft center of gravity and the $x0z$ plane in symmetry plan of body. This system is often used as an inertial reference system, because it is rather close to the inertial system (when the main axes are the inertial axes of the body). The wind reference system is used for studying the flight mechanics and flight performance. This system is derived from the body system by directing the x axis to the aircraft real motion velocity (**Figure 2a**). (The $x0z$ plane is still in aircraft symmetry plan.) **Figure 2b** applies the body axis to aerofoil (wing section) 2D case.

The first explanation of the lift generation can be derived using complex potential flow. Applying the double source and uniform flow for modeling the fixed cylinder moving in ideal (viscosity less) flow, the results show that no lift and no drag are generated on the body, and the velocity/pressure/distribution on the cylinder is symmetric (**Figure 3a**). By including the potential vortex into the model being described, the rotated cylinder in the ideal flow, the results lead to fundamental theorem, called Kutta-Joukowski theorem explaining that the lift is generated because the vortex appears around the body (**Figure 3b**). In ideal flow, there is no drag (flow in **Figure 3b** is symmetric to vertical axis).

Kutta-Joukowski theorem [6]: $L = \rho V\Gamma$, where L is the lift; ρ and V are air density and velocity; and Γ is the vorticity, which means lift can be generated only in cases when a vortex appears around the body.

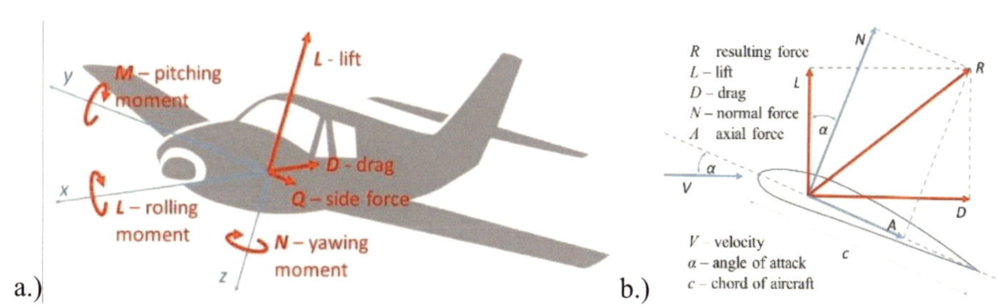

Figure 2. The components of the aerodynamic force and moments generated on the aircraft (a) and airfoil (b).

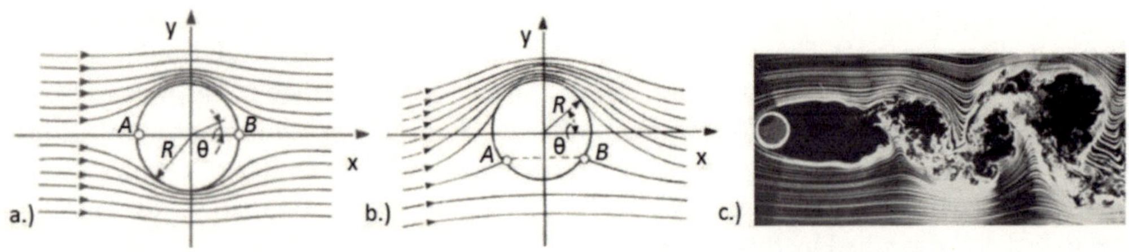

Figure 3. The flow around the fixed (a) and rotated (b) cylinders in ideal flow and flow separation from cylinder moving real flow [16] (c) (*R* is the radius of the cylinders [17]; *A* and *B* are the stagnation points in which the flow velocity equals to zero).

In real flow, because the viscosity, drag is generated too (D'Alambert paradoxon), due to flow separation (**Figure 3c**). Prandtl introduced an excellent idea [18]: flow near the body surface must be described as real flow, and flow outside this layer, called boundary layer, can be represented as ideal flow. In the boundary layer, the flow might be laminar, when the sublayers near the body surface move parallel, but with different velocities, or turbulence, when the flow particles move in a chaotic ways [1, 5, 6]. The developed boundary layer theories [19, 20] may well define the skin friction drag, drag appearing in boundary layer (**Figure 4a**).

On the other hand, the drag has several components [1–8] (**Figure 4b**). The induced drag is affected by the vortex lines separating at the wing tips. There is no lift without vortex, while vortex induces some drag.

The drag resulting from pressure distribution on the body surface and skin friction drag together is called as profile drag. Flow separation drag is the drag initiated by separation of flow (at high speed or at high angle of attack). Wave drag is caused by the shock wave system appearing at high subsonic, transonic, and supersonic speeds. The interference drag is the extra drag affected by interaction of the flows around the different elements of aircraft (or even different aircrafts). The 3D drag is an interesting special drag component caused by effects of 3D aspects. Finally, the aircraft components like radio antenna add the aircraft components' drag. Often, especially for subsonic cruise speed, the drag is classified by the use of so-called causal breakdown, and when the pressure and friction drag are composed from flat plate friction, drag components are affected by protuberances, roughness, and incremental profiles.

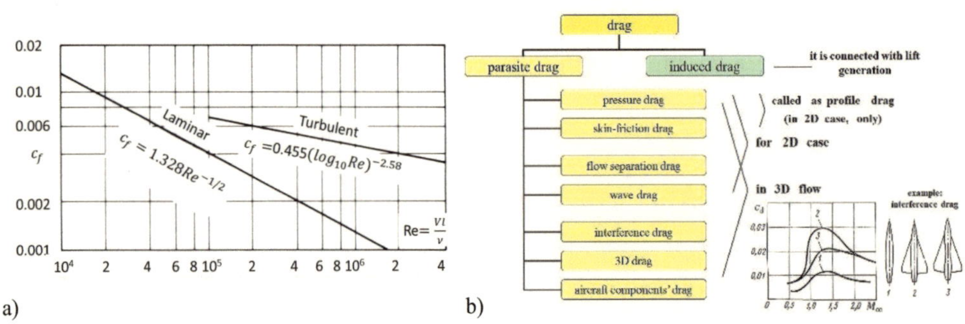

Figure 4. Skin friction drag of the thin plate (a) depending on the Reynolds number (Re = flow velocity × length/air kinematic viscosity) and classification of the drag (b).

Mathematical investigation and calculation of the aerodynamic forces and moments are supported by computational fluid dynamics (CFD) [21–23]. Nowadays, several well-applicable software are available. The cost- and time-effective CFD technology allows to simulate and compute (i) all the desired quantities (stream functions and vorticity, including the integral quantities as lift, drag, and moments), (ii) with high resolution in space and time and it is applicable to (iii) actual flow domain, (iv) virtual problems, and (v) realistic operating conditions, as well as (vi) excellent visualization and (vii) systematic data analysis of the results (**Figure 5**). Numerical aerodynamics may give excellent results in simplified cases or after serious adaption (verification and validation) to the investigated situations. Generally, the quality of the CFD results depends on managing the uncertainties (real turbulence and their modeling [26]) and so-called unacknowledged errors (as logical mistake in using the software, errors in parametrization, models of boundary conditions, bugs, etc.).

The practical measurements and estimations of the aerodynamic forces, moments, and their coefficients by use of wind tunnel and flight tests comparing to CFD are very costly and require lot of time (up to several years) [27–29]. The practical methods might be used for study (i) in limited number of quantity, (ii) in limited number of operational points and time instant, (iii) in limited range of problems and operating conditions, and as usual (iv) with use of small-scale models (**Figure 6a**) or specially equipped aircraft.

The practical measurements (including the flow visualization, studying the flow separation, developing the streamlined bodies, too) support the (i) understanding of the flow structure, (ii) measuring and identification of the aerodynamic coefficients, (iii) verification and validation of

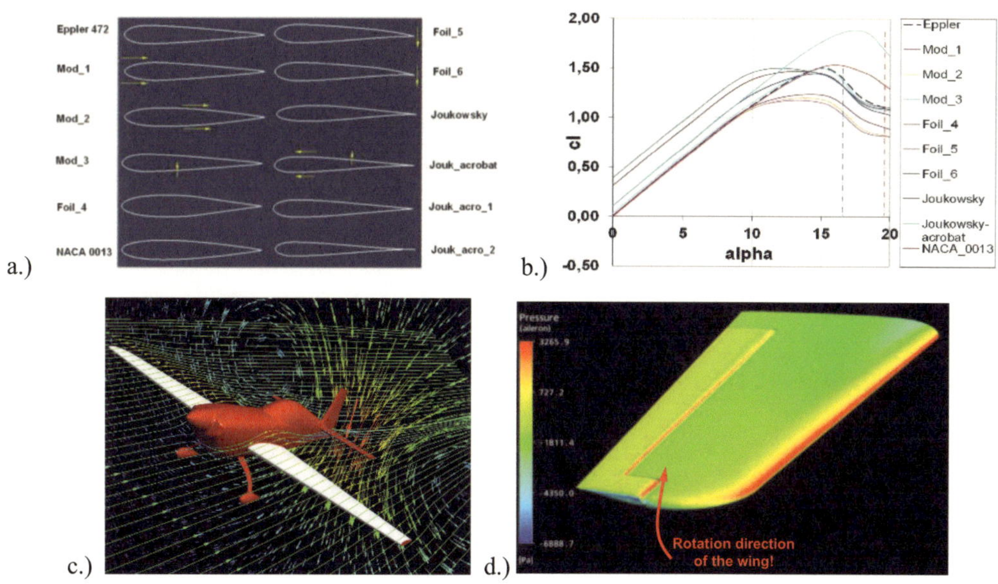

Figure 5. The typical example of CFD application to development of a special acrobatic aircraft Corvus Racer 540 [24, 25]. (a) Series of investigated profiles modified from Eppler 472 and Joukowsky, (the yellow arrows show the ways of modification), (b) determined lift coefficient angle of attack curves, (c) the optimized fuselage and the 3D flow with tip vortex, and (d) pressure distribution on wing in high speed rolling.

a.) 

b.) 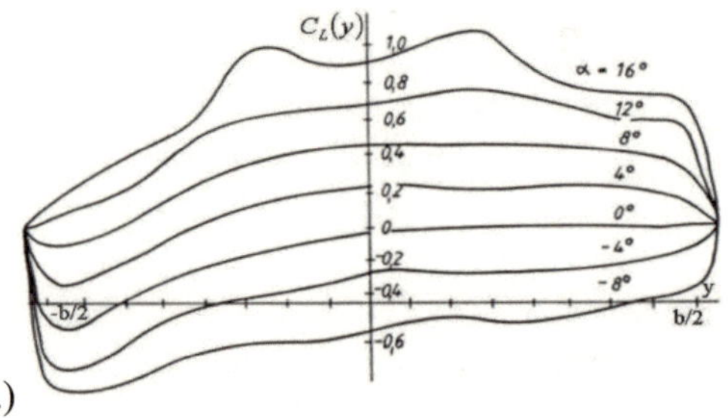

Figure 6. The practical measurements: (a) An-225 Mria and space shuttle group model in the wind tunnel at the TsAGI [30], (b) lift coefficient distribution along the wing span (b/2) of deformed (left side) and nondeformed (right hand) wing [31].

the CFD methods, (iv) optimizing shape for cruise flight mode, and (v) studying the most dangerous flight mode, aircraft approach and landing.

Figure 6b shows how the real lift distribution depends on the flight conditions, namely how the deformation of wing deformed under loads has influence on the actual lift distribution.

Generally, the differences in calculated and measured wind tunnel lift coefficient reach 7–8%, while, for example, the differences between the measured wind tunnel and flight test drag coefficient equal to 5–10% and up to 18% at the transition period from subsonic to supersonic flights [32]. During the periodic angle of attack oscillation of the wing, there is a large hysteresis in the lift coefficient—angle of attack function. So, there are considerable differences in steady and unsteady regime.

These thoughts on aerodynamic force and moment generation demonstrate that the theoretical calculation and the practical measurements cannot independently provide full and correct description for aerodynamic forces and moments. At first, the semiempirical methods were developed and applied for aircraft aerodynamic design and calculation of the aerodynamic characteristics [33–37]. Later, with gaining in prestige of CFD, the role of modeling of aerodynamic coefficient increased.

3. Aerodynamic coefficients

The motion of aircraft can be described by a system of equations describing the motion of center of gravity of aircraft and its rotation around it [13–15]. The general form of the system contains stochastic, partial nonlinear differential equations with delays.

So, this motion can be defined by solving the inertial equations that should be coupled with the equations describing the aerodynamic (gas dynamic) and elastic phenomena (see central part of **Figure 1**) [9, 11, 12]. The latter equations are coupled through the aircraft shape and structure. The time-dependent aerodynamic equations describe the instantaneous aerodynamic effects on the aircraft assumed in the form of aerodynamic forces and moments depending on the state of the flow field surrounding the aircraft, the motion variables, the aircraft shape deformation, and the initial conditions.

In the first approximation, the instantaneous aerodynamic force depends on air density, ρ, and velocity, V, and mean geometrical parameter of the body, namely wing span, s. That by use of methods of dimensional analysis can be represented in the form:

$$F = C\rho^{\alpha} V^{\beta} s^{\gamma}, \tag{1}$$

where C is the coefficient. The exponents α, β, and γ must be derived from the condition that the dimension of the different sides of equation should be equal. Using the results $\alpha = 1$, $\beta = 2$, and $\gamma = 2$, the aerodynamic force can be calculated as:

$$F = C\rho^1 V^2 s^2 = 2CAR\frac{\rho V^2}{2}\frac{s^2}{AR} = C_F\frac{\rho V^2}{2}S, \tag{2}$$

where c_F is the so-called nondimensional force coefficient, AR is the aspect ratio ($AR = s^2/S$), and S is the wing area. From here, the aerodynamic force and moment coefficients are:

$$C_F = \frac{F}{\frac{\rho V^2}{2}S}, \qquad C_M = \frac{M}{\frac{\rho V^2}{2}Sc_a}. \tag{3}$$

Here C_a is the aerodynamic chord.

The aerodynamic forces and moments, as well as their aerodynamic coefficients, can be represented by their components:

$$C_F = \left[C_x(=C_D), C_y, C_z(=C_L)\right]^T, \qquad C_M = \left[C_l, C_m, C_n\right]^T, \tag{4}$$

according to the axes of the applied reference system (here wind system) [1, 14].

The nondimensional aerodynamic coefficients fully describe the aircraft aerodynamics [1–8]. The basic aerodynamic characteristics of airfoils [38, 39] are shown in **Figure 7**. The left first figure shows the typical changes in lift coefficient with increase in angle of attack that begins with linear function, followed by nonlinear form at high angle of attack and dropping after separating the flow from the upper surface of aerofoil at the so-called critical angle of attack.

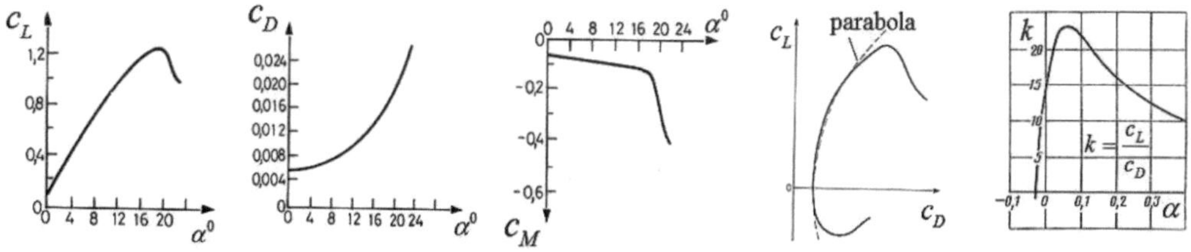

Figure 7. Aerodynamic characteristics of an airfoil: lift, drag, and moment coefficients as function of angle of attack, polar curve, and "goodness" factor.

This phenomenon is called stall. The drag coefficient increases with growing lift coefficient due to the induced drag. The moment coefficient follows the changes in lift and drag coefficients, especially after stall. The polar curve ($C_L = f(C_D)$) and "goodness" factor ($k = (C_L/C_D = f(\propto))$) explain the relatively low angle of attack and must be realized during the most important flight regime, during the cruise flight for having minimum drag, minimum required thrust, and minimum fuel consumption.

As it is well known, the subsonic and supersonic aerodynamics is principally different. The "classic" airfoils with blunt leading edge cannot be applied, because their drag tends to the infinity nearing to Mach number (velocity related to the sound speed in the same condition) equals to one. At the supersonic speed only, the airfoils (wing and fuselage) with sharp leading edge can be applied (**Figure 8a**).

Figure 8b demonstrates how changes in some parameters may radically affect the aerodynamic characteristics. In case of high aspect ratio wing, the vortex generating the lift as a vortex tube along the wing span separates at the wing tips and causes the induced drag. The low aspect ratio delta wing has unique aerodynamic picture. The flow separating from the wing leading edge and the caused by this separated flow vortexes moving back on the top of the wing generate extra lift at high angle of attack, and the stall appears at 45–70°, only.

The aerodynamic coefficients depending on the flight modes and flight maneuvers are managed by use of control surfaces and motion devices as flaps, slots, and generally by all the devices deviating and changing the geometry like undercarriage system, braking parachutes, etc. [1–8]. For example, the flaps at wing trailing edge and slat at the leading edge (making slots between the slat and mean wing) are used for increasing the lift (and drag) allowing to reduce the take-off and landing speeds for making safer flight modes. The flaps increase the lift coefficients ("moving" the lift coefficient and angle of attack curve left and up in **Figure 9a**), while the slats/slots increase the critical angle of attack, (because they do not change the airfoil/wing chamber). In **Figure 9a**, at low lift coefficient region, the effect of a special leading edge flap, called Krueger flap, is shown, too.

The flaps are deflected on the lower angle during take-off than during landing, because they increase the drag, too. **Figure 9b** demonstrates these changes in polar curve diagrams depending on the flap deflection.

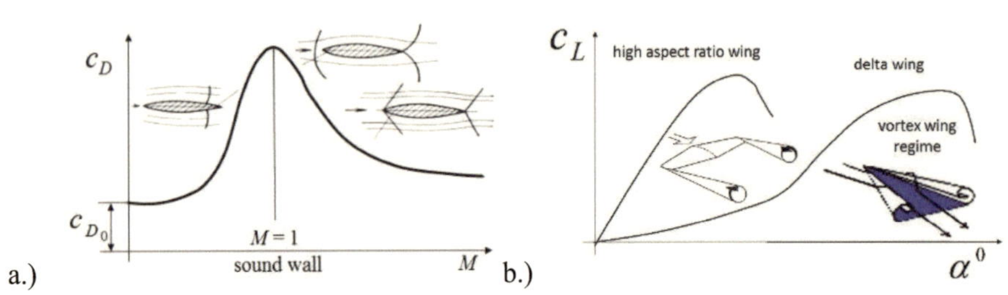

Figure 8. Further specific aspects: (a) drag coefficient of the supersonic airfoil depending on the Mach number and (b) lift coefficients generated on the high and low aspect ratio (delta) wings.

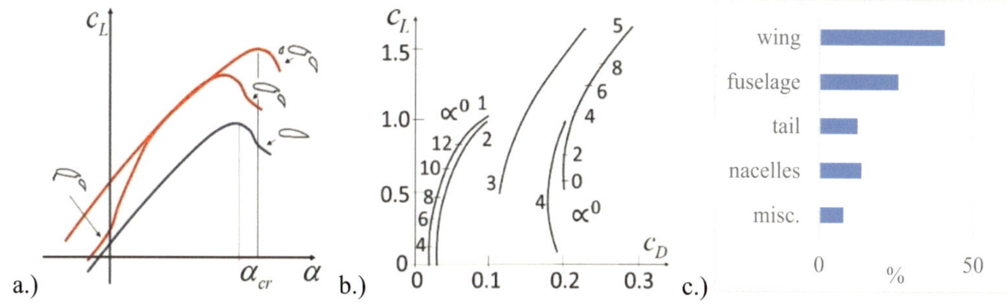

Figure 9. An integrated aerodynamic characteristic of the wing (airfoil) and flap/slats (a), polar curve diagrams of a middle size passenger aircraft (b), and drag coefficient breakdown (c). (In figure b, the numbers at curves depict the flight conditions as 1. cruise flight (all devices are closed), 2. undercarriage system is open, only, 3. take-off regime (flaps deflected near 30^0), 4. landing condition (flaps deflected up to 45°), 5. all the wing mechanisms are opened (slats and interceptors).

Figure 9c calls attention to the final aerodynamic coefficients that always are composed from the coefficient generated on/by the aircraft elements.

These examples underline that the aerodynamic characteristics depend on the state of the flow field surrounding the aircraft, like air viscosity, motion variables, e.g., linear and angular velocities, real geometrical characteristics reflecting the effect of the deflection of the control elements and the deformation of aircraft, and they may have a sensitive dependence on the initial conditions (**Figure 1**). Therefore, the aerodynamic coefficients are given in the form of functions of different variables, like position angles and velocities of aircraft, flow characteristics, namely Reynolds number, Mach number (speed), deflection angles of aerodynamic control surfaces, control forces, etc. [1–8]. These functions are very nonlinear and very complicated. In case of dynamic changes in basic parameters (like angle of attack) and especially in case of oscillation motion, the aerodynamic coefficients contain the hysteresis-type nonlinearities depending on the frequencies and amplitudes of oscillation. So, different simplified, more complex, and special models and mathematical representations are needed.

4. The first (simple) aerodynamic models

The mathematical descriptions of the aerodynamic coefficients are called as aerodynamic models [1–8]. First models were based on the work of Bryan [40], who used two principal assumptions: the aerodynamic forces and moments depend only on the instantaneous values of the motion variables, and their dependence is of linear character. Therefore, the simple models of the aerodynamic coefficients can be expanded into a Taylor series about the reference states.

$$C_A(t) = C_{A_0} + \sum_{i=1}^{n} C_{a_{p_i}} p_i(t), \tag{5}$$

where C_A is the aerodynamic coefficient, p_i, $i = 1, 2, \ldots, n$ are the parameters, C_{A_0} is the aerodynamic coefficient at $p_i = 0, \forall i$ and the $C_{a_{p_i}}$ is the partial derivative coefficient.

$$C_{A_{p_i}} = \left(\partial C_A / \partial p_i\right)_{p_i=0}. \tag{6}$$

For example, the pitching moment in simplified case can be represented by the following term:

$$C_m(t) = C_{m_0} + C_{m_V} V(t) + C_{m_q} q(t) \tag{7}$$

where the C_{m_V} and C_{m_q} are the moment coefficient derivatives:

$$C_{m_V} = (\partial C_m / \partial V)_{V=0}, \qquad C_{m_q} = (\partial C_m / \partial q)_{q=0}. \tag{8}$$

Later, taking into account the more realistic characteristics of the nonsteady flow associated with the aircraft motion, the results received refused both assumptions of Bryan. The new models introduced by Glauert [41] contain additional elements taking into consideration the effect of the past history of the aircraft motion on the current aerodynamic forces and moments [42–44]. The flight dynamic, stability, and control had been applied to Glauert's idea in more general form. The aerodynamic coefficients were defined by the use of c linear air reaction theory outlined by Etkin [41, 45]. In this approach, the coefficients are linearized around the predefined operational points. The interactions between the angle of attack, the control surface deflection, and aerodynamic coefficient, as well as the time lag effect on the aerodynamics, were taken into account. The aerodynamic model was rewritten in form, like the following model of the lift coefficient:

$$\begin{aligned} C_L(t) = \; & C_{L_0} + C_{L_\alpha}\alpha(t) + C_{L_{\alpha^2}}\alpha^2(t) && + C_{L\dot{\alpha}}\dot{\alpha}(t) && + C_{L_\delta}\delta(t) \\ & \textit{linear part} \quad + \textit{non} - \textit{linear part} && + \textit{time lag} && + \textit{control effect}. \end{aligned} \tag{9}$$

Here $C_{L\dot{\alpha}} = (\partial C_L / \partial \dot{\alpha})_{\dot{\alpha}=\dot{\alpha}_0}$ is the derivative of the lift coefficient, respectively, to the rate of change in angle of attack $\dot{\alpha} = \partial \dot{\alpha} / \partial t$ and it represents the time lag addition determined by using the assumption that it is proportional to $\dot{\alpha}$. The partial derivatives of the aerodynamic models are used as stability derivatives [13–15, 42, 45, 46] and they must be multiplied by changes in variables $(\Delta\alpha, \Delta\alpha^2, \ldots)$ as deviations from the flight regime (operational point) at which the coefficients are determined. The derivatives should be independent. Principally, quantities α and $\dot{\alpha}$ are not independent. So the models like (9) approximate the aerodynamic coefficient in the form of a mathematically incorrect expansion [11].

5. Classic aerodynamic models

The simplified aerodynamic coefficient representations adapted to the real situations and real problems today are the widely and most used aerodynamic models. The different types of simple classic aerodynamic models [1–8, 13–15, 42–46] are shown in **Table 1**.

The usual linearized formulations of the aerodynamic models and nonlinear models described above can only be used for detailed investigations where the aircraft motion is prescribed. This is the mean difficulty with such models.

Coefficients	Models	Remarks
$C_L(t)$	$C_{L_0} + C_{L_\alpha}\alpha(t) + C_{L_{\delta_e}}\delta_e(t)$ $C_{L_0} + C_{L_\alpha}\alpha(t) + C_{L_{\alpha^2}}\alpha^2(t) + C_{L_{\delta_e}}\delta_e(t)$ $C_{L_0} + C_{L_\alpha}\alpha(t) + C_{L_{\alpha^2}}\alpha^2(t) + C_{L_{\dot{\alpha}}}\dot{\alpha}(t) + C_{L_{\delta_e}}\delta_e(t) + C_{L_{\dot{\delta}_e}}\dot{\delta}_e(t)$ $C_{L_0} + C_{L_\alpha}\alpha(t) + C_{L_{\alpha^2}}\alpha^2(t) + C_{L_{\dot{\alpha}}}\dot{\alpha}(t) + C_{L_{\delta_e}}\delta_e(t) + C_{L_{\dot{\delta}_e}}\dot{\delta}_e(t)$ $+C_{L_{\alpha\delta_e}}\alpha\delta_e(t) + C_{L_{\dot{\alpha}\delta_e}}\dot{\alpha}\delta_e(t) + \ldots$	Simple, linearized model, simple, nonlinearized model (longitudinal motion), model used for nonsteady state aerodynamics, full model
$C_D(t)$	$C_{D_0} + C_{D_V}V(t) + C_{D_q}q(t) + C_{D_\alpha}\alpha(t) + C_{D_{\delta_e}}\delta_e(t)$ $C_{D_0} + C_{D_V}V(t) + C_{D_q}q(t) + C_{D_\alpha}\alpha(t) + C_{D_{\alpha^2}}\alpha^2 + C_{D_{\dot{\alpha}}}\dot{\alpha}(t)$ $+C_{D_{\delta_e}}\delta_e(t) + C_{L_{\dot{\delta}_e}}\dot{\delta}_e(t)$ $C_{D_0} + C_{D_V}V(t) + C_{D_q}q(t) + C_{D_\alpha}\alpha(t) + C_{D_{\alpha^2}}\alpha^2 + C_{D_{\dot{\alpha}}}\dot{\alpha}(t)$ $+C_{D_{\delta_e}}\delta_e(t) + C_{L_{\dot{\delta}_e}}\dot{\delta}_e(t) + C_{D_{\alpha\delta_e}}\alpha\delta_e(t)$ $+C_{D_{\dot{\alpha}\delta_e}}\dot{\alpha}\delta_e(t) + C_{D_{\delta_r}}\delta_r(t) + \ldots$ $C_{D_0} + C_{D_\alpha}\alpha(t) + C_{D_{\alpha^2}}\alpha^2(t) + C_{D_{\alpha^4}}\alpha^4(t) + C_{D_\beta}\beta(t) + C_{D_q}q(t)$	Simple, linearized model, simple, nonlinearized model, full model, special model for high angle of attack situations
$C_m(t)$	$C_{m_0} + C_{m_V}V(t) + C_{m_q}q(t) + C_{m_\alpha}\alpha(t) + C_{m_{\delta_e}}\delta_e(t)$ $C_{m_0} + C_{D_M}M(t) + C_{D_{M^2}}M^2 + C_{m_q}q(t) + C_{m_{\alpha M}}\alpha M(t) + C_{m_{\alpha^2}}\alpha^2(t)$ $+C_{m_{\delta_e}}\delta_e(t) + C_{m_{\varphi_s}}\varphi_s(t) + C_1\frac{\Delta x_{cg}}{C_a} + C_2\frac{\Delta y_{cg}}{C_a}$ $C_{m_0} + C_{m_\alpha}\alpha(t) + C_{m_{\alpha^2}}\alpha^2(t) + C_{m_{\alpha^3}}\alpha^3(t) + C_{m_\beta}\beta(t) + C_{m_q}q(t)$ $+C_{m_{\delta_e}}\delta_e(t)$ $C_{m_0} + C_{m_\alpha}\alpha(t) + C_{m_{\alpha^2}}\alpha^2(t) + C_{m_{\beta^2}}\beta^2(t) + +C_{m_q}q(t) + C_{m_{\delta_e}}\delta_e(t)$ $+C_{m_{\alpha\delta_e}}\alpha\delta_e(t)$	Simple model, model applied to study the required fuel consumption, models for investigation of the nonlinear effects

Here δ_e and δ_r are the deflection angle of the control surface elevator and rudder, q is the dynamic pressure, β is the sideslip angle, i.e., angle between the x axes of the body and wind reference systems, and Δx_{cg} and Δy_{cg} are the coordinates of deviated position of the center of gravity.)
Remarks. Often small c is used instead of capital C in aerodynamic coefficients. Sometimes m_z is applied instead of C_m.

Table 1. Different aerodynamic models and their possible applications.

The full aerodynamic description of the aircraft requires a lot of component models. These models are often defined as semiempirical models, such models are based on theoretical bases, adapted to measured data. **Figure 10a** shows an example for use of such methods developed for aircraft aerodynamic design. The derivative coefficient of the lift generated on the nose part of a fuselage depends on the flight Mach number (M_∞) and ratio of lengths of central and nose parts of the fuselage.

Another example is the calculation of the fuselage friction drag coefficient appearing at zero angle of attack:

$$C_{D_{0,f}} = C_f\eta_t\eta_M\frac{S_{fus_{wet}}}{S_{fus_M}} \tag{10}$$

where C_f is the skin friction drag coefficient thin plate, η_t *and* η_M are coefficients taking into account the effect from body thickness and flow velocity, $S_{fus_{wet}}$ *and* S_{fus_M} are the fuselage

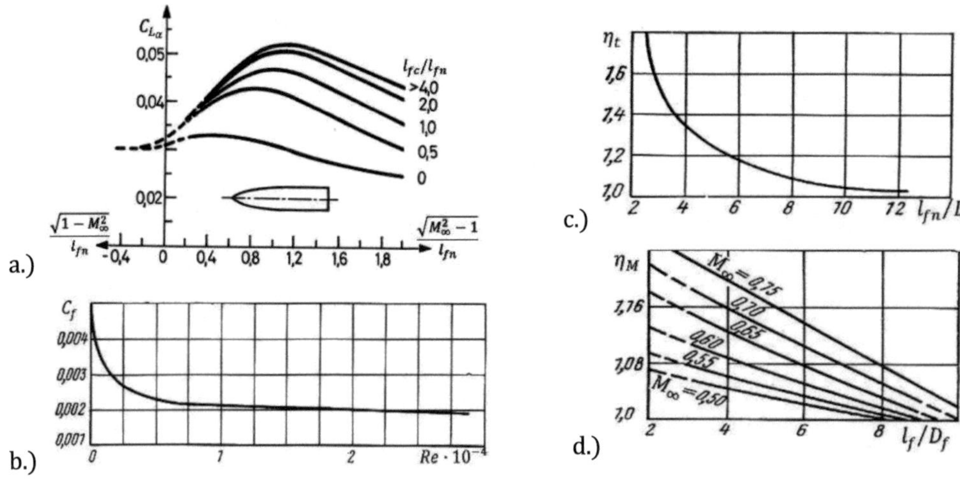

Figure 10. Practical figures supporting the estimation and evaluation [5] of the aerodynamic coefficients: (a) lift slop coefficient for estimation of the lift generated on the fuselage nose, (b)–(d) figures defining the estimation of the friction drag generating on the fuselage.

so-called surface wetted area and area of the fuselage mean (maximum) cross-section area, and $D_f, l_f,$ *and* l_{fn} are the mean diameter, length of fuselage, and length of the nose section of the fuselage (in **Figure 10**). The $C_f, \eta_t,$ *and* η_M coefficients can be estimated from **Figure 10b–d**, while the fuselage wetted area can be calculated with the use of the following formulas:

$$S_{fus_{wet}} = \pi D_f l_f \left(1 - \frac{2D_f}{l_f}\right)^{2/3} \left(1 - \frac{2D_f^2}{l_f^2}\right) \textit{if the } {}^{l_f}/{}_{D_f} \geq 4.5 \text{ or } S_{fus_{wet}} = 2.53 D_f l_f. \tag{11}$$

As it had been outlined already, the first simplified models were adapted to the wide flight dynamics, stability, and control investigations and to the different form of aircraft [13–15, 42–46]. The modern control introduced the state space representation of the linearized system of equations describing the aircraft spatial motion:

$$\dot{\mathbf{x}} = \mathbf{A}\mathbf{x} + \mathbf{B}\mathbf{u} \tag{12}$$

x and **u** are the state and control (input) vectors, while **A** and **B** are the state and control matrices. In simplified case, when the aircraft is modeled as rigid body, the state vector contains the components of the linear and rotational (angular) velocities – $\mathbf{x} = [u, v, w, p, q, r]^T$. The control vector is composed of control inputs including the control surfaces' deflection, deflections of other moving elements as flaps, slat, as well as the changes in trust: $\mathbf{u} = \left[\delta_e, \delta_r, \delta_a, \delta_f, \delta_s, \ldots, n_T\right]^T$. Here the control elements are the deflection angle of elevator, rudder, aileron, flaps, slats, and engine revolution speed. Principally, because of the linearization, the state and control vectors contain the changes in velocity components and deflection angles related to the operational (initial) condition. Because of symmetry, the motion equations can be divided into two subsystems: longitudinal and lateral motion. The state vector of the longitudinal motion model (motion of aircraft in the vertical plane, only) contains the u, w, q and additionally the pitch (or climb) angle, θ. The **A** and **B** elements are special derivative coefficients.

The aircraft longitudinal motion can be modeled by

$$\begin{aligned} m\frac{du}{dt} &= Tcos(\alpha+\varphi_T) - D - W\,sin\theta \\ m\frac{dw}{dt} &= Tsin(\alpha+\varphi_T) + L - W\,cos\theta\,. \\ I_y\frac{dq}{dt} &= M \end{aligned} \tag{13}$$

equations that are defined in body system of reference. Here m and W are the aircraft mass and weight, T is the trust and φ_T is the engine built angle, angle between the trust direction, x_b is the axis of the body system of reference, and I_y is the inertia moment component. Supposing the $cos\,(\alpha+\varphi_T)\approx 1$, $sin\,(\alpha+\varphi_T)\approx 0$ and taking into account the X, Z, and M are the components of the total forces and moment component due to x, z, and y axes, respectively, Eq. (13) can be rewritten into the space state representation form:

$$\begin{bmatrix} \dot{u} \\ \dot{w} \\ \dot{q} \\ \dot{\theta} \end{bmatrix} = \begin{bmatrix} \frac{X_u}{m} & \frac{X_w}{m} & -gcos\theta_0 & 0 \\ \frac{Z_u}{m} & \frac{Z_w}{m} & -gsin\theta_0 & 0 \\ \frac{M_u}{I_y} & \frac{M_w}{I_y} & \frac{M_q}{I_y} & 0 \\ 0 & 0 & 1 & 0 \end{bmatrix} \begin{bmatrix} u \\ w \\ q \\ \theta \end{bmatrix} + \begin{bmatrix} \frac{X_{\delta_e}}{m} & \frac{X_{n_T}}{m} \\ \frac{Z_{\delta_e}}{m} & \frac{Z_{n_T}}{m} \\ \frac{M_{\delta_e}}{I_y} & \frac{M_{n_T}}{I_y} \\ 0 & 0 \end{bmatrix} \begin{bmatrix} \delta_e \\ n_T \end{bmatrix}. \tag{14}$$

Here, the components $X_u, X_w, \ldots, M_u, \ldots, M_q$, are called stability derivatives and $X_{\delta_e}, \ldots, M_{n_T}$ are the control derivatives. Of course, the aerodynamic total forces and moments might be estimated by the sum of the derivatives of the force and moment components relevant to the given state and control vector elements.

For instance, X_u should be determined from the first equation of (13):

$$\begin{aligned} m\frac{du}{dt} &= T(V,\Omega,\delta_e,n_T) - C_D\frac{\rho V^2}{2}S - W\,sin\theta \\ mX_u &= \frac{\partial T}{\partial u} - \frac{\rho V_0^2}{2}\frac{\partial C_D}{\partial u} - C_D\frac{\rho S}{2}\frac{\partial V^2}{\partial u}, \end{aligned} \tag{15}$$

and in simple case, when $V = u$, the dimension-less derivative equals to:

$$X_u = \frac{T_u}{\frac{\rho V_0 S}{2}} - V_0 C_{D_V} - 2C_D. \tag{16}$$

The drag coefficient, C_D, can be represented by the models described earlier.

The static and dynamic stability, flight dynamics (as maneuvers, maneuverability, departure to the critical regimes, and recovery from there) and control design, and control synthesis are required to know the aerodynamic characteristics of the aircraft elements and aircraft devices, too. For instant, the hinge moment coefficient (m) of the control surfaces (elevator, rudder, and ailerons) can be represented by the following simplified models:

$$\begin{aligned} m_e &= m_{e_\alpha}\alpha + m_{e_{\delta_e}}\delta_e + m_{e_{\delta_e T}}\delta_{eT} \\ m_r &= m_{r_\beta}\beta + m_{r_{\delta_r}}\delta_r + m_{r_{\delta_r T}}\delta_{rT} \quad , \\ m_a &= m_{a_\alpha}\alpha + m_{a_p}p + m_{a_{\delta_a}}\delta_a + m_{a_{aT}}\delta_{aT} \end{aligned} \tag{17}$$

where index T depicts the trim tabs and the p is the pitch rate.

Finally, another excellent example demonstrates the interaction between the different theories. As **Figure 8b** shows, the lift coefficient on the delta wing depends on the vortex generated at the leading edge. Polhamus [47] created and explained a special formula for lift coefficient calculation:

$$C_{L_{\Delta w}} = K_p \ sin \ \alpha \ cos^2\alpha + K_V \ cos \ \alpha \ sin^2\alpha \tag{18}$$

Here the first part comes from the small angle of attack potential lifting surface theory. The K_p is the lift curve slop, $sin \ \alpha$ accounts for true boundary condition, and the $cos^2\alpha$ arises from the Kutta-type condition at the leading edge. In second part of the formula (18), the $K_V \, sin^2\alpha$ gives the potential flow leading edge suction, i.e., vortex normal force, and the $cos \ \alpha$ defines its component in the lift direction.

The classic models are well applied in identifying them from flight data and developing the flight simulation methods, too [29, 48].

6. Developed aerodynamic models

The classic aerodynamic models cannot be applied to accurate description of the aircraft motion at high angle of attack, aircraft maneuvers, dynamic, oscillation motion or aerodynamic characteristics in flutter, etc. Tobak [49] introduced a model structure. He made a special assumption: the changes in aerodynamic coefficients are linear functions of changes in variables that are independent of the past history of these variables, namely on all values that these variables have taken over the course of the motion prior to time τ. For example, the change in pitching moment can be defined by following functions:

$$\Delta C_m = \frac{\Delta C_m(t-\tau)}{\Delta\delta}\Delta\delta + \frac{\Delta C_m(t-\tau)}{\Delta(ql/V)}\Delta(ql/V) \tag{19}$$

Here δ is the motion of aircraft along the z axis in body axis system ($\delta = z$), q is the angular velocity around the y axis, and the derivatives depend on elapsed time t -τ rather than on t and τ.

The derivatives in Eq. (19) come from solution of linear equation of gas dynamics. However, the linearity assumption does not rest on the assertion that change in pitching moment, (ΔC_m), is linear dependent on changes in variables, $\Delta\delta(=\Delta\delta_e)$ and $\Delta(ql/V)$. So, these two increments must not be linear additives in Eq. (19).

Principally, the aerodynamic pitching moment coefficient response to variations δ and q. These variations can be broken into a large number of small step changes (**Figure 11**). No matter how large the values of δ and q at the beginning of steps, the derivatives depend on the t -τ, only. The limits of these functions

$$\lim_{\Delta\delta\to 0}\frac{\Delta C_m(t-\tau)}{\Delta\delta}=C_{m_\delta}(t-\tau),\qquad \lim_{\Delta(ql/V)\to 0}\frac{\Delta C_m(t-\tau)}{\Delta(ql/V)}=C_{m_q}(t-\tau) \tag{20}$$

are called as the linear indicial pitching moment responses per unit step changes in δ and ql/V, respectively [11, 49].

Using this indicial function concept to calculate the aerodynamic coefficients, Tobak [11, 49] replaced Bryan's function with a linear functional in the form of the linear superposition integral like:

$$C_m(t)=C_m(0)+\int_0^t C_{m_\delta}(t-\tau)\frac{d}{d\tau}\delta(t)\mathrm{d}\tau+\frac{1}{V}\int_0^t C_{m_q}(t-\tau)\frac{d}{d\tau}q(t)\mathrm{d}\tau. \tag{21}$$

In reality, the functions of aerodynamic coefficient and derivatives depend on all the past values of the motion variables. In accordance to Volterra's description, the aerodynamic coefficient as function can be given in the form of a functional:

$$C_m(t)=\mathrm{G}[\delta(\xi),q(\xi)] \tag{22}$$

Generally, the whole time/past history of motion variables is unknown. Therefore, the functional (22) can be replaced by a functional describing the dependence on the past in the form of analytical functions in the neighborhood of $\xi=\tau$ reconstructed from the Taylor series expansions of the coefficients about $\xi=\tau$. This obtains for example:

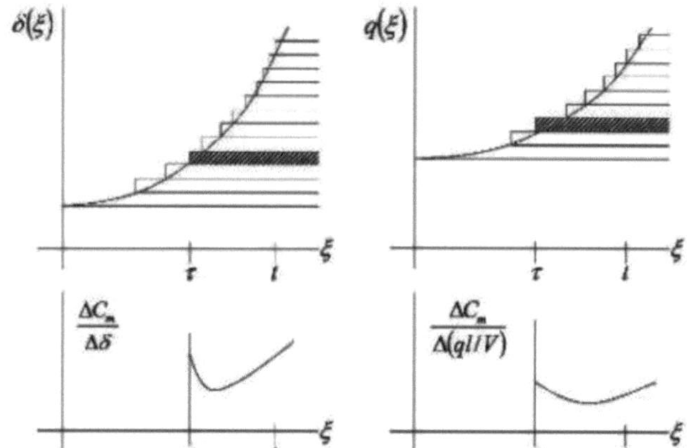

Figure 11. Simulation of incremental responses.

$$C_{m_\delta}[\delta(\xi), q(\xi); t, \tau] = C_{m_\delta}\left(t, \tau; \delta(\tau), \dot{\delta}(\tau), \ldots, q(\tau), \dot{q}(\tau), \ldots\right). \tag{23}$$

Hence, at most, only the first few coefficients of expansions of $\delta(\xi)$ and $q(\xi)$ need be retained to characterize correctly the most recent past [49], which is all the indicial response remembers. Using the two coefficients of $\delta(\xi)$, for example, implies matching the true past history of δ in magnitude and slope at the origin of the step, thereby approximating $\delta(\xi)$ by a linear function of time

$$\delta(\xi) \approx \delta(\tau) - \dot{\delta}(\tau)(\tau - \xi). \tag{24}$$

With application of this approach. Eq. (21) can be rewritten into the following form:

$$\begin{aligned} C_m(t) = C_m(0) &+ \int_0^t C_{m_\delta}\left(t, \tau; \delta(\tau), \dot{\delta}(\tau), q(\tau), \dot{q}(\tau)\right) \frac{d}{d\tau} \delta(t) \mathrm{d}\tau \\ &+ \frac{1}{V} \int_0^t C_{m_q}\left(t, \tau; \delta(\tau), \dot{\delta}(\tau), q(\tau), \dot{q}(\tau)\right) \frac{d}{d\tau} q(t) \mathrm{d}\tau \end{aligned} \tag{25}$$

This method of model definition is more attractive then (21) and gives the possibility of taking into account the considerable nonlinearities, time lag, and hysteresis, too. All the developed models follow from this model formation. For example, in case of slowly varying motion, Eq. (25) may be formalized in a more general form,

$$C_m(t) = C_m(0) + \int_0^t C_{m_\delta}(t - \tau; \delta(\tau), q(\tau)) \frac{d}{d\tau} \delta(t) \mathrm{d}\tau + \frac{1}{V} \int_0^t C_{m_q}(t - \tau; \delta(\tau), q(\tau)) \frac{d}{d\tau} q(t) \mathrm{d}\tau \tag{26}$$

still capable of embracing a fairly broad range of nonlinear problems of aerodynamics.

The use of indicial aerodynamic functions is a rather complex task even for 2D [50].

The next step in developing the aerodynamic models was made by Goman and his colleague [51, 52]. They had formulated the aerodynamic coefficient models in the form of a state space representation:

$$C_a = C_a(\boldsymbol{\xi}(t)\boldsymbol{\eta}(t)), \tag{27}$$

where

$$\dot{\boldsymbol{\eta}}(t) = \mathbf{g}\left(\boldsymbol{\eta}(t)\boldsymbol{\xi}(t)\dot{\boldsymbol{\xi}}(t)\right) \tag{28}$$

and

$$\boldsymbol{\xi}(t) = \left[\boldsymbol{x}(t)^T \boldsymbol{u}(t)^T\right]^T. \tag{29}$$

Here **η** is an internal additional state vector and **x** and **u** are the state and control vectors from the aircraft motion models (see Eq. (12)). For instance, Ref. [51] described the aircraft longitudinal dynamics by introducing the internal state variable representing the vortex burst point location along the chord of a triangular wing.

7. Advanced aerodynamic models

The collection of large databases of practical wind tunnel and flight test measurements and wide use of rapidly developing methods of computational fluid dynamics and a series of new methods have developed for modeling the aerodynamic coefficients. Three different approaches can be applied: (i) approximation and interpolation, (ii) analytical models and special models, and (iii) models developed using soft computing models.

The polynomial, and trigonometric interpolation, spline or regression models can be used for determining the aerodynamic coefficients or aerodynamic forces directly. For example, reference [53] uses the Lagrange interpolation to determine the lift and drag coefficient when studying the takeoff taxiing. The piecewise cubic Hermite interpolating polynomial and Spline are applied [54] to calculating the derivative of the pressure distribution on airfoil for determining the laminar-to-turbulent transition. The transition is identified as the location of maximum curvature in the pressure distribution. The Chebyshev polynomials and their orthogonality properties were applied [55] for approximation of unsteady generalized aerodynamic forces from the frequency domain into the Laplace domain, acting on a Fly-By-Wire aircraft. The results were compared with Padé method and validated on the aircraft test model.

The oscillation in changes of the aerodynamic forces and their coefficients contributes to the most interesting areas of developing the aerodynamic coefficient models. This area has two major parts: (i) oscillation of the aircraft elements, like flutter, and (ii) oscillation flight of aircraft. The first and today valued as fundamental studies were published in 1920s and 1930s. Wagner [56] dealt with unsteady lift on airfoil due to abrupt changes in angle of attack and he calculated the circulation around the airfoil in response to a step in angle of attack. Theodorsen [57] extending the Wagner concept developed a model for quasi-steady thin airfoil theory including added-mass forces and the effect of wake vorticity.

$$C_L = \pi\left[\ddot{h} + \dot{\alpha} - a\ddot{\alpha}\right] + 2\pi\left[\alpha + h + \alpha\left(\frac{1}{2} - a\right)\right]C(k) \tag{30}$$

Here the added-mass force taken into account by the first addend, while the second one defines the quasi-steady lift from thin airfoil theory by a transfer function C(k) as lift attenuation by the wake vorticity. The *h* is the vertical position of airfoil, *a* is the pitch axis with respect to 1/2 chord, and the Theodorsen's transfer function C(k) is expressed in terms of Hankel functions:

$$C_L(k) = \frac{H_1^{(2)}(k)}{H_1^{(2)}(k) + iH_0^{(2)}(k)}, \tag{31}$$

where $H_n^{(2)}(k) = J_n - Y_n, \quad n = 0, 1$ are Bessel function, and $k = \omega c/2V_\infty$, where ω is the motion frequency, c is the airfoil chord, and V_∞ is the free stream velocity. This approach is well applicable nowadays, too (see [58, 59]).

In 1980s and 1990s during the development of the supermanoeuvrable and thrust vectored aircraft, the hysteresis in aerodynamic coefficient was intensively studied. These aircrafts fly at critical regimes, near or at the border of the flight envelopes. Thrust vectored aircraft uses the controlled poststall flights.

The hysteresis effects in aerodynamic coefficients can appear in different forms depending on the oscillation frequency [60–63]. **Figure 12** shows typical hysteresis caused by stall in normal force coefficient at the high angle of attack flight [64] and in steady-state pitching moment response [65].

The considerable nonlinearities in the aerodynamic coefficients that generate the hysteresis in aerodynamic characteristics near the critical angle of attack in stall and poststall domain, of course, are well investigated by practical methods in wind tunnels [66–69].

The formation of flow separation at the critical angle of attack is a quite complete process [70], and the hysteresis [64] shown in **Figure 12a** fundamentally depends on the frequency of changes in the angle of attack. Therefore, the approximation of these characteristics is a difficult task. The models described earlier cannot ensure the required accuracy in the full region of parameter variations. The aerodynamic models used in the early works were based on fitting polynomials [71] or cubic [72, 73] or bi-cubic [74, 75] splines as interpolation schemes for measured data given in the form of table. In some cases [76], the methods that worked out for bifurcation analysis did not require further smoothing and the linear interpolation had been applied.

In many cases, the aerodynamic coefficients are given in table form [68, 69] or directly estimated from the flight tests [29, 77]. Data can be obtained by special analytical models [78]:

$$C_F = b_0 + \sum_{i=1}^{n} b_i arctan((\alpha - c_i)d_i), \tag{32}$$

where b_0, b_i, c_i are the constants.

a.) b.)

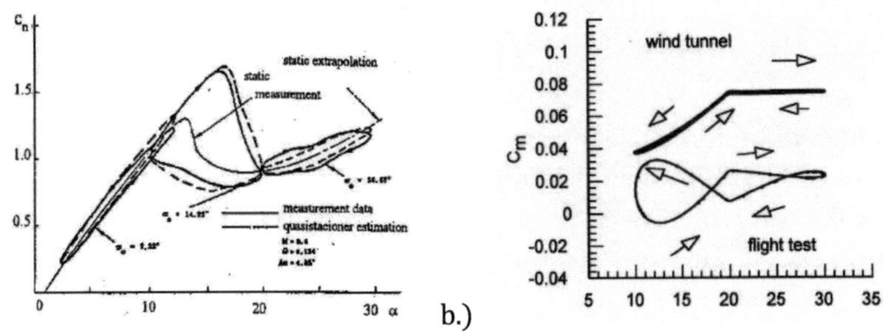

Figure 12. Typical hysteresis in aerodynamic coefficients. (a) At high angle of attack [64], (b) moment coefficient estimated from the wind tunnel and flight test of F-16XL-1 [65] (at reduced frequency $k = 0.054$).

This model was developed especially for the approximation [78] of experimental data received from wind tunnel investigations [68, 69]. The aerodynamic models obtained in form (32) can be used in full AoA region from −10 to 90°. Analytical models of type (32) have a great advantage; namely, there is no α value, where the derivative of this function does not exist.

Figure 13 shows some examples of developed analytical models defined for different speeds and elevator deflections with linear approximation between them. One example of these NASA-backed representation of the actual derivative involves four to eight arcus tangent functions:

$$C_{m\dot{\alpha}} = -\frac{0.02}{\pi} arctg\left(-5\pi\frac{\alpha-1}{18}\right) + 0.5\, arctg(5(\alpha-6)) - 0.8 arctg\left(\frac{\alpha-18}{2}\right) + 0.9 arctg\left(\frac{\alpha-45}{2}\right) - 0.9 \qquad (33)$$

Since 1990s, by developing numerical aerodynamics, and applying the methods of soft computing, new types of aerodynamic coefficient representations have been developed. It seems the most applied method is based on using the neural network [79, 80]. The other papers predicted the aerodynamic coefficient of transport aircraft with the use of artificial neural networks [81], simulated the dynamic effects of canard aircraft aerodynamics [82], used genetic algorithm optimized neural networks for predicting the practical measurements [83], determined the global aerodynamic modeling with multivariable spline [84], and applied the fuzzy logic modeling to the aircraft model identification [85] and nonlinear unsteady aerodynamics [86]. Principally all the numerical methods might be applied. For instance, the aircraft stability and control can be modeled with the use of wavelet transforms [87] or even the computed stability derivatives can be applied directly in aerodynamic shape optimization [88]. Nowadays, the computer capacity and sizes allow to use the real-time on-board identification of the nonlinear aerodynamic models [89].

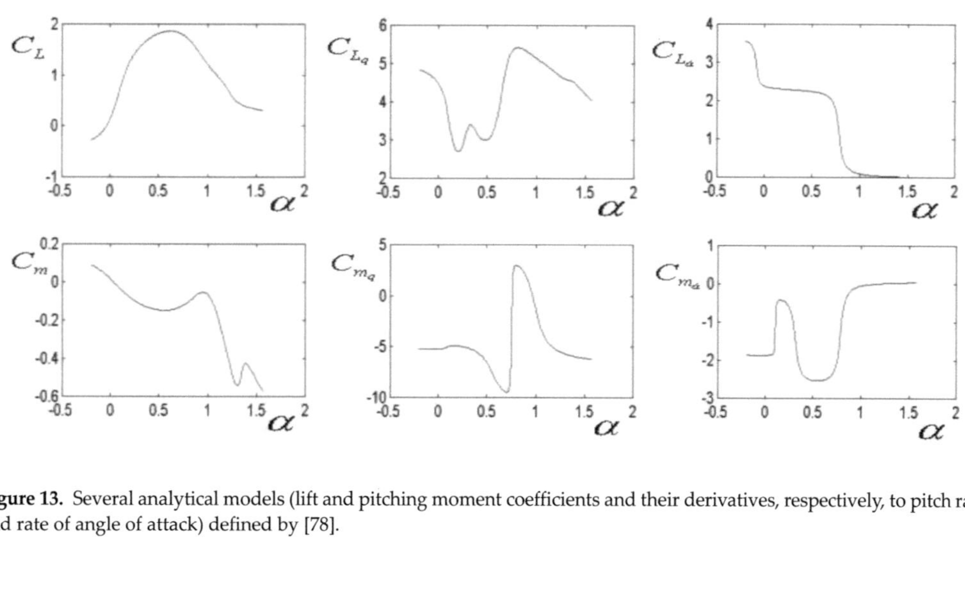

Figure 13. Several analytical models (lift and pitching moment coefficients and their derivatives, respectively, to pitch rate and rate of angle of attack) defined by [78].

Two specific aspects must be underlined: (i) the computational fluid dynamics may easily determine the aerodynamic coefficients by integration of the calculated surface pressure distribution and (ii) all the aerodynamic coefficient models described earlier can be applied, while better using the models as simple as possible depending on the goal and object of their application.

8. Applicability of the aerodynamic coefficient models

Table 1 has given already some advises for possible application of the different models of the aerodynamic coefficients. The developed and advanced models open new fields of application including the investigation of the fully nonlinear situations including the aircraft chaotic motions and provide more accurate derivatives for maintaining stability and control.

The model application is based on the known technology identification, evaluation, and selection methodology [90, 91]. This methodology can be adapted to the aerodynamic coefficient model selection by using the following major steps:

1. Definition of the object, objectives, and goals

   Define the object as thrust vectored aircraft, wing flutter, and objectives like managing the thrust vectored aircraft poststall motion, or reducing the amplitudes of the wing oscillation motion. Derive the goals from the objectives as investigation, design the new system, and control or manage with the object, etc.

2. Identification of the applicable models

   Derive the preliminary specification of the required models for the definition of the object, objectives, and goals. Namely, the models might be local (used locally to a part of the aircraft or to well-defined motion regime, like cruise flight, only) or global (applying to the whole aircraft, or to the large part of flight envelope). Estimate which nonlinearities, delay, and hysteresis in aerodynamic coefficient may appear that should be taken into account.

   Identify the possible models from literature review, preliminary investigations, brainstorming, etc.

3. Evaluation of the identified models

   Study the identified models: especially evaluate how they can be integrated into the existing or planned systems (compatibility), how their changes or modification may improve their applicability for supporting the objectives (apply the morphological matrix), how their deployments have impact on the applying systems (impact matrix—effect on the solutions like using the different control philosophy and control technique), how effective, safe, sustainable, etc. is their application, and how they might have influence on their selection (decision matrix).

   The evaluation must be dealt with development of the final systems, including the production, supply chain, market introduction, etc.

The candidate models might be tested in simulation, or even in laboratory or flight tests. The tests must cover the full range of possible flight regimes and situations, and the result must be evaluated against the predefined indicators. The sensitivity analysis may detect the most important parts or elements of the models.

4. Selection of the best models for the aerodynamic coefficients required for reaching the predefined objectives

   The selected models must be as simple as possible, while their application is (life cycle) cost-effective and they must support the objectives.

5. Development of the systems applying the selected aerodynamic coefficient models

   The system developments include the hardware and software developments and a study of the total impact (effect on the life cycle cost, safety, security, and environment as chemical emissions and noise) and verification and validation of the created systems.

6. Final decision and deployment

   Depending on the previous points, the identification, evaluation, and selection process might be finished or started from the beginning. Of course, with the changes in aircraft structures, new ways of operation, application of the new solutions, and new emerging technologies, the aerodynamic coefficient models always must be refined or even the identification, evaluation, and selection process must be repeated again and again followed by improving or developing new solutions and systems improving the aircraft aerodynamic shape, aerodynamic characteristics, performance, stability, disturbed motion, and controllability.

Table 2 gives some advises on how to use the different models of the aerodynamic coefficients.

Models	Typical examples	Applicability
Simple	$C_L = C_{L_0} + C_{L_\alpha}\alpha(t) + C_{L_{\delta_e}}\delta_e(t)C_D = C_{D_0} + C_{D_\alpha}\alpha(t) + C_{D_{\alpha^2}}\alpha^2$	Local models for simplified cases like drag-required thrust-fuel consumption for cruise flight, studying the linearized static motion static stability
Classic	$C_L = C_{L_0} + C_{L_\alpha}\alpha(t) + C_{L_{\alpha^2}}\alpha^2(t) + C_{L\dot{\alpha}}\dot{\alpha}(t) + C_{L_{\delta_e}}\delta_e(t) + C_{L_{\dot{\delta}_e}}\dot{\delta}_e(t)$	Static, queasy static models for full range of flight envelope including the high angle of attack flights, linearized motion equations, semiempirical models defining the stability and control derivatives, basic unsteady models
Advanced	$C_m(t) = C_m(0) + \int_0^t C_{m_\delta}(t-\tau)\frac{\mathrm{d}}{\mathrm{d}\tau}\delta(t)\mathrm{d}\tau + \frac{1}{V}\int_0^t C_{m_q}(t-\tau)\frac{\mathrm{d}}{\mathrm{d}\tau}q(t)\mathrm{d}\tau$	Models for dynamic motion, analysis of the critical flights, study and control of dynamic effects including delays, hysteresis in models, etc., developing the global models, critical flights
Developed	$C_F = b_0 + \sum_{i=1}^{n} b_i \arctan((\alpha - c_i)d_i),$	Models for all specific flight situation, and regimes by approximation of the available wind tunnel and/or flight tests measured data, developing models by use of soft computing based on classic or developed models

Table 2. Some recommendations on the usage of the aerodynamic coefficient models.

9. Example of use of an advanced aerodynamic coefficient model

The Department of Aeronautics, Naval architecture and Railway Vehicles at the Budapest University of Technology and Economics (operating two flight simulators, one air traffic management laboratory with several working environment for ATCOs, small gas turbines, water channel, etc.) is active in computational fluid dynamics [92, 93], vehicle design [24, 25, 94], vehicle motion simulation [95, 96], developing original and radically new technologies [97–99], and has worked on investigation of the thrust vectored aircraft motion at high angle of attack in poststall domain [100, 101], approximation of the motion after stall [102], and unconventional and critical flights [103, 104]. One of the excellent applications of the analytical models of the aerodynamic coefficients is their using in bifurcation analysis of the poststall motion of thrust vectored aircraft.

Only the longitudinal motion was investigated. The applied system of equations defined by the use of body axis was reduced to four dimensions given in the following form [101]:

$$\begin{aligned}
\dot{u} &= -qw + \frac{X}{M} - gsin\theta + \frac{T_x}{M} \\
\dot{w} &= -qu + \frac{Z}{M} - gcos\theta + \frac{T_z}{M} \\
\dot{q} &= \frac{C_m \overline{q} S c_A + Xl_z + Zl_x - T_x L_{xe}}{I_y} \\
\dot{\theta} &= q
\end{aligned} \tag{34}$$

where $X = \overline{q}S(C_L sin\alpha - C_D cos\alpha)$, $Z = \overline{q}S(C_L cos\alpha + C_D sin\alpha)$, $C_L = C_{L_0} + \frac{c_A}{2V}\left(C_{L\dot{\alpha}}\dot{\alpha} + C_{L_q}q\right)$, $C_D = C_{D_0} + \frac{c_A}{2V}\left(C_{D\dot{\alpha}}\dot{\alpha} + C_{D_q}q\right)$, $C_m = C_{m_0} + \frac{c_A}{2V}\left(C_{m\dot{\alpha}}\dot{\alpha} + C_{m_q}q\right)$, $T_x = Tcos\delta_{vp}$, $T_x = Tsin\delta_{vp}$. Here X and Z are the force components to x and z axis, M is the pitching moment, $\overline{q}$ are q are the dynamic pressure and pitch rate, respectively, and T, T_x, and T_z are the thrust and its components.

Different types of simple and classic aerodynamic coefficient models were applied that could not result in stable and acceptable solutions. Therefore, the described system of equations and analytical models of aerodynamic coefficients were filled up by data of F/A-18 aircraft [68, 69, 78]. These models defined the hysteresis effects, as well, and they may be used in full region of the possible changes in angle of attack (see **Figure 13**).

The system of equation was solved by different numerical methods (Runge-Kutta and Adams-Moulton) with different step size. Software MATHLAB and ACSL were used in the simulations. The results received were stable and the same at time steps 10^{-2} and 10^{-6} s.

Figure 14 shows the equilibrium surface obtained in the thrust-thrust deflection parameter space (left side) and the bifurcation curves (right side) for flight regime V = 0.3 M and H = 15,000 ft. (T = 22.7 kN, δ_{vp} = 0°).

As it can be seen, the poststall domain of the thrust vectored aircraft motion can be divided into six different subspaces. The subspaces are divided by bifurcations. There were found two

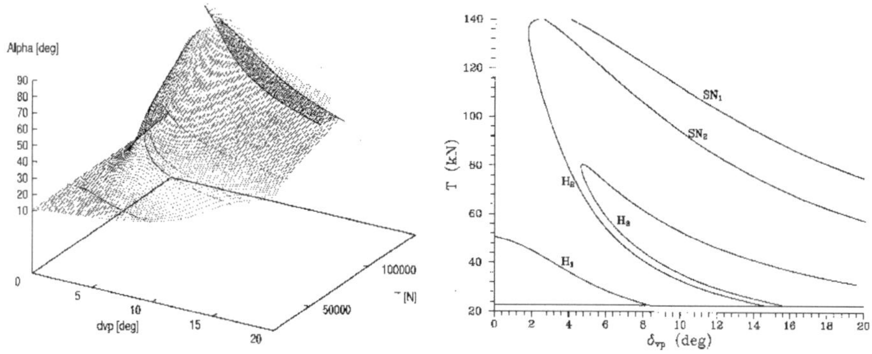

Figure 14. Equilibrium surface on the thrust-thrust deflection parameter space (left side) and the bifurcation curves (right side).

different types of bifurcation, e.g., Hopf (H) and saddle-nodes (SN) bifurcations. The first region at the small thrust and small angle of thrust-deflection is characterized the phugoid motion of aircraft before the stall. Oscillation of speed is greater than changes in angle of attack. The fighter slowly returns to the stable position.

As chosen by increasing the thrust and thrust-deflection, the system reads the first Hopf bifurcation (H_1) (a small amplitude limit cycle appears at the bifurcation point). Further by increasing thrust and thrust deflection, there is no stable state of the aircraft. Over this second region, changes in the thrust and thrust deflection cause lack of stability before and poststall oscillation of the aircraft. This oscillation tends to the limit cycle and the angle of attack can reach the 90°.

By another Hopf-bifurcation, the system gains back its stability in the poststall regimes. This is the narrow streak area inside the second zone. At high thrust and thrust-deflection, the saddle-node bifurcation (SN) emerges creating jump phenomena. The motion of aircraft in zone appearing after first saddle-node bifurcation curve is an oscillation motion in the poststall domain.

Finally, in the last zone at very high thrust and thrust-deflection, an overpulling appears, when the angle of attack reaches over 90° during the first period of motion after changes in the thrust or thrust deflection.

The bifurcations were followed by continuation method. The input was generated in the thrust deflection (not in the thrust), as it would have been usual nonlinear approach. Components T_x, T_z were computed by the following formulas:

$$T_x = T cos(\delta_{vp} + \varepsilon cos\omega t), \quad T_z = T sin(\delta_{vp} + \varepsilon cos\omega t). \tag{35}$$

In some cases, several interesting changes were found in angle of attack response on oscillation in thrust deflection (**Figure 15**). Little bit nicer representation of this chaotic changes in angle of attack is given in **Figure 16**. This is a 3D phase plot by redrawing of simulation results shown in **Figure 15**. Such phase plot represents the chaos in system output received as results of periodic excitations and it is called as chaotic attractors.

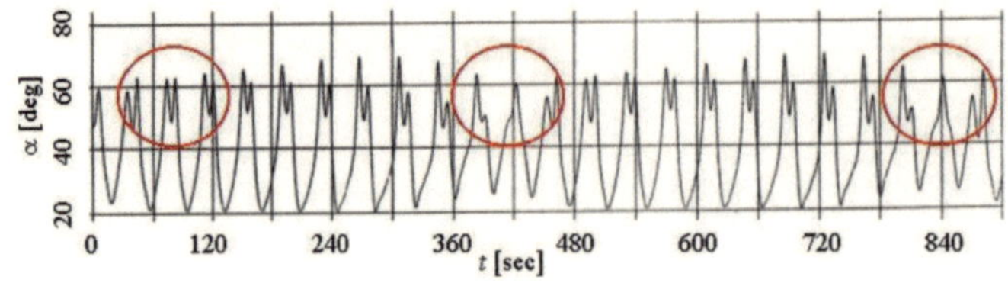

Figure 15. Angle of attack response initiated by thrust oscillation with amplitude 2° and frequency 0.33 rad/s applied to initial condition of equilibrium at $T = 35$ kN and $\delta_{vp}=10^0$.

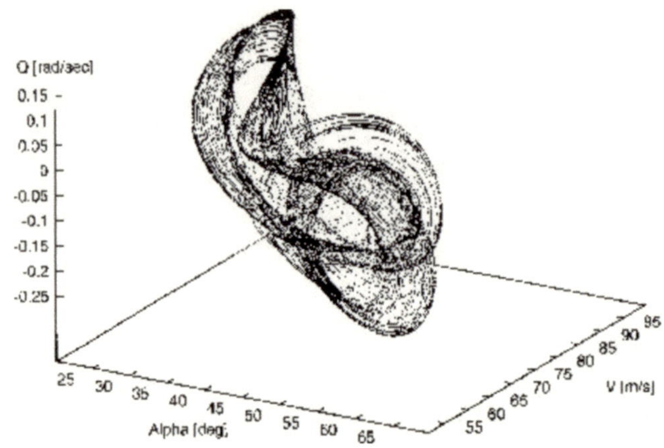

Figure 16. 3D phase plot of chaotic attractor is described by simulation results given in **Figure 15**.

A small change in system parameters or in excitations can cause a relatively big change in system output (**Figure 17**). For example, reduction of excitation frequency from 0.33 to 0.32 rad/s involved reduction of chaotic behavior in response and resulted in periodic orbits (see **Figure 17**). In some cases, the periodic orbits are reduced to one (it may be strange) limit cycle. The other figure shows that around 0.9 rad/s another type of noninear phenomenon appears, which is called period doubling bifurcation. At this point, the time period becomes twice as long (no sudden catastrophic change). Decreasing the frequency, a cascade of period doubling bifurcation happens leading to chaos around 0.65 rad/s. **Figure 17** demonstrates several chaotic regions can appear (see chaotic window at the $\omega = 0.35$ rad/s in **Figure 17**).

Further investigation of the aerodynamic coefficient models had been studied by use of sensitivity analysis and changes in structure of the models. The sensitivity analysis had shown that the changes in aerodynamic derivatives for 5 or 1% did not result in considerable changes in response on the applied oscillated thrust deflection.

Using the same mathematical model, initial condition, and excitation (at $T = 35$ kN, $\delta_{vp} = 10$ deg., $\Delta\delta_{vp} = 2$ d° and $\omega = 0.33$ rad/s), the simulations were realized with the use of different aerodynamic coefficient models, in which different parts, or derivatives, were omitted. The results show that elements cause the changes in angle of attack responses (**Table 3**).

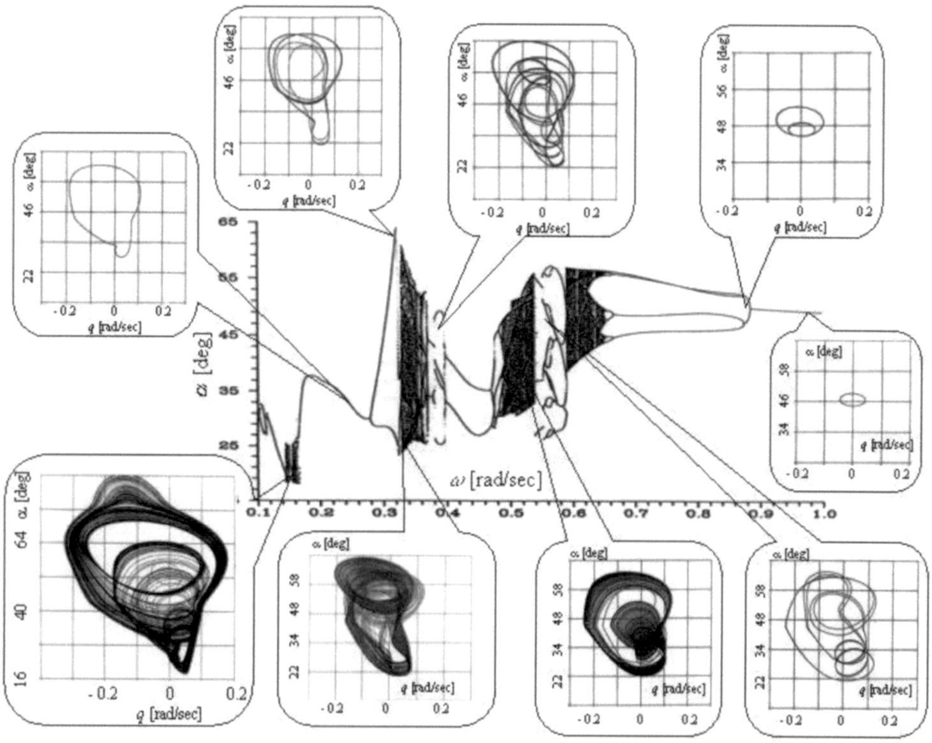

Figure 17. Stroboscopic map and different phase plots demonstrate the results of simulation thrust oscillation with amplitude 2° with varied frequency applied to initial condition of equilibrium at $T = 35$ kN and $\delta_{vp}=10^0$.

C_{M_q}	$C_{M\dot{\alpha}}$	C_{L_q}	$C_{L\dot{\alpha}}$	Remark	C_{M_q}	$C_{M\dot{\alpha}}$	C_{L_q}	$C_{L\dot{\alpha}}$	Remark
+	+	+	+	Chaos	—	+	+	+	Limit cycle
+	+	+	—	Chaos	—	+	+	—	Limit cycle
+	+	—	+	Chaos	—	+	—	+	Limit cycle
+	+	—	—	Chaos	—	+	—	—	Limit cycle
+	—	+	+	Transient chaos	—	—	+	+	Limit cycle
+	—	+	—	Transient chaos	—	—	+	—	Limit cycle
+	—	—	+	Transient chaos	—	—	—	+	Limit cycle
+	—	—	—	Transient chaos	—	—	—	—	Limit cycle

Table 3. Influence of aerodynamic model structure on the aircraft poststall motion initiated by cosine excitation in thrust deflection angle (sign shows the omitted elements).

10. Conclusions

Aerodynamics deals with interaction of air and bodies moving in it. The major task of aerodynamics is to define and describe the aerodynamic forces and moments generated on the bodies.

Because of the very complex ways of causing the aerodynamic forces and moments, the nondimensional aerodynamic force and moment coefficient and series of their models had been developed for the last hundred years. This short chapter tries to show the different aspects having influences on "burning" the aerodynamic forces and moments and their contributing elements.

The aerodynamic coefficient models can be classified as simple, classic, developed, and advanced models. The models use the partial derivative coefficients, indicial step responses, analytical models, interpolation and approximation of the available wind tunnel, flight test, or numerical simulation data, and models are generated by utilization of the soft computing methods.

There is no unique and well-applicable method to selecting the required and best coefficient models. Always the object- and goal-oriented models must be selected. The identification, evaluation, and selection process may use the general methodology: (i) definition of the object, objectives, and goals, (ii) identification of the applicable models, (iii) evaluation of the identified models, (iv) selection of the best models, (v) development of the systems applying the selected aerodynamic coefficient models (including the verification and validation, too), and (vi) final decision.

There are some recommendations supporting the selection of the aerodynamic coefficient models and an example demonstrates using a special model to complex motion of thrust vectored aircraft in poststall domain.

Author details

Jozsef Rohacs

Address all correspondence to: jrohacs@rea-tech.eu

Budapest University of Technology and Economics, Budapest, Hungary

References

[1] Anderson DJ Jr. Fundamentals of Aerodynamics. St. Louise, New York: McGraw-Hill Inc.; 1991. 792 pp

[2] Houghton EL, Carpenter PW. Aerodynamics for Engineering Students. Oxford, Amsterdam: Butterworth Heinemann; 2003. 614 pp

[3] Schlichting H, Truckenbrodt E. Aerodynamic des Flugzeugs. 3rd ed. Berlin, Heidelberg, New York: Springer-Verlag; 2001. Teil 1 480 pp, Teil 2 515 pp

[4] Krasnov NF. Aerodinamika (in Russian). 2nd ed. Moscow: Vysshaya Skola; 1976. T1 383 pp, T2 468 pp

[5] Mhitaryan AM. Aeriodinamika (in Russian). Masinostroyeniye: Moscow; 1976. 446 pp

[6] McCormick BW. Aerodynamics, Aeronautics and Flight Mechanics. 2nd ed. New York: John Wiley and Sons., Inc.; 1995. 652 pp

[7] Rohacs J, Gausz ZS, Gausz T. Aerodinamika (in Hungarian). Typotex: Budapest; 2012. 220 pp

[8] McLean D. Understanding Aerodynamics: Arguing from the Real Physics. West Sussex, England: John Wiley and Sons, Inc.; 2013, 550 pp

[9] Rohács J, Gránásy P. Effects of non-linearities in aerodynamic coefficient on aircraft longitudinal motion. In: Sivasundaram S, editor. Non-Linear Problems in Aviation and Aerospace. New York: Gordon and Breach; 2000, pp. 281-296

[10] Rohács J. Bifurcation analysis of aircraft poststall motion. In: Proceedings of the 7th Mini Conference on Vehicle System Dynamics, Identification and Anomalies; Budapest, 2000; Department of Railway Vehicles, BUTE, Budapest. 2001. pp. 53–76

[11] Tobak M, Schiff LB. Aerodynamic Mathematical Modelling—Basic Concepts, Dynamic Stability Parameters. AGARD-LS-114. 1981. pp. 1.1–1.31

[12] Bisplinghoff RL, Ashley H, Halfman RL. Aeroelasticity. Cambridge, MA: Addison-Wesley Publishing Company, Inc.; 1955. 860 pp

[13] Etkin B, Reid LD. Dynamics of Flight, Stability and Control. 3rd ed. New York: John Wiley and Sons, Inc; 1996, 383 pp

[14] Cook MV. Flight Dynamics Principles. London: Arnold; 1997, 352 pp

[15] Stengel R. Flight Dynamics. Princeton University Press; 2004

[16] Szabo J, editor. Aviation lexicon (in Hungarian: Repülési lexikon). Budapest: Akademiai Kiadó; 1991. T.1 and T.2, 623 + 603 pp

[17] Van Dyke M. An Album of Fluid Motion. Stanford: The Parabolic Press; 1982. 174 pp

[18] Prandtl L. Über Flüssigkeitsbewegung bei sehr kleiner Reibung. In: Krazer A, editor. "Verhandlungen des Dritten Internationlen Mathematiker-Kongresses in Heidelberg; August 8–11, 1904; Leipzig: Druck und Verlag von B. G. Teubner; 1905. pp. 484-491

[19] Karman T. Über laminare und turbulente Reibung. ZAMM Zeitschrift für Angewandte Mathematik und Mechanik. 1921;**1**(4):232-252

[20] Schlichting H. Grenzschicht-Theorie. Kalsruhe: Verlag G. Braun; 1965. 736 pp. (Boundary Layer Theory, 7th ed. New York: McGraw-Hill Book Company, Inc.; 1979. 817 pp. [in English])

[21] Versteeg HK, Malalasekera, W. An Introduction to Computational Fluid Dynamics, The Finite Volume Method. 2nd ed. Harlow, London: Pearson, Prentice Hall; 2007. p. 517

[22] Cummings RM, Mason WH, Morton SA, McDaniel DR. Applied Computational Aerodynamics, A Modern Engineering Approach. Cambridge University Press; 2015. 849 p

[23] Charlott JJ, Hafez MM. Aerodynamics: And Related Numerical Methods. Dordrecht, Heidelberg: Springer; 2015. 620 pp

[24] Rohacs J, Voloscsuk A, Gecse T, Ovari Gy. Innovation process management for reducing the time to market. Aerospace—The Global Industry; November 2–4, 2010; Exhibition Centre Frankfurt, Main Germany, Conference Proceedings; AIRTECH GmbH and Co. KG. 2010. p. 21. ISBN: 978-3-942939-00-3

[25] Rohacs D, Farkas Cs, Ovari Gy, Rohacs J. Time and cost minimized integrated product and production process development for customized low volume advanced technology. In: Repüléstudományi Közlemények, XXIX; 2017. No. 3

[26] Nicols. Turbulence Models and Their Application to Complex Flows. 4.01 version. onlines guide. https://overflow.larc.nasa.gov/files/2014/06/Turbulence_Guide_v4.01.pdf (downloaded at 16 of July, 2017)

[27] Krasnov NF. Prikladnaya Aerodinamika (Practical Aerodynamics—in Russian). Moscow: Vysshaya Skola; 1975. 731 p

[28] Kimberlin RD. Flight Testing of Fixed Wing Aircraft. AIAA; 2003. 435 p

[29] Klein V, Morelli EA. Aircraft System Identification—Theory and Practice. AIAA Education Series; 2006. 499 p

[30] Central Aerohydrodynamic Institute. Photo gallery. Downloaded at 16 of July, 2017; http://tsagi.com/pressroom/photogallery/

[31] Martynov AK. Prikladnaya aerodinamika [Practical Aerodynamics – in Russian]. Moscow: Vysshaya Skola; 1972. 448 p

[32] Saltzman EJ, Ayers TG. Review of flight-to-wind-tunnel drag correlation. Journal of Aircraft. 1982;**10**:801-811

[33] Torenbeek E. Synthesis of Subsonic Airplane Design. Delft: Delft University Press; 1976. 598 p

[34] Obert E. Aerodynamic Design of Transport Aircraft. Delft: Delft University Press, IOS Press; 2009. 656 p

[35] Küchemann D. The Aerodynamic Design of Aircraft. Oxford, New York: Pergamon Press; 1978. 564 p

[36] Roskam J. Airplane Design Part VI: Preliminary Calculation of Aerodynamic Thrust and Power Characteristics. Sixth Printing. Lawrance, Kansas, US: Dar Cooperation; 2017. 584 p

[37] Sobester A, Forrester AIJ. Aircraft Aerodynamic Design: Geometry and Optimization. John Wiley and Sons; 2014. 262 p

[38] Abott IA, Doenhoff AE, Stivers LS Jr. Summary of Airfoil Data. NACA Report No. 824; 1945. 261 p. https://ntrs.nasa.gov/archive/nasa/casi.ntrs.nasa.gov/19930090976.pdf [visited at 12 July 2017]

[39] UIUC Airfoil Coordinates Database. UIUC Applied Aerodynamics Group, Department of Aerospace. http://m-selig.ae.illinois.edu/ads/coord_database.html [visited at 12 July 2017]

[40] Bryan GH. Stability in Aviation. Macmillan & Co.; 1911. 192 p

[41] Cowley WL, Glauert H. Effect of the lag of the Donwash on the longitudinal stability of an airplane and on the rotary derivative Mq. ARC R&M 718; 1921

[42] Горбатенко СА, Макашов ЭМ, Полушкин ЮФ, Шефтель ЛВ. Механика полета: (Общие сведения. Уравнения движения). Инженерный справочник, Машиностроение, Москва; 1969. 420 с

[43] Белоцерковский С, Скрипач Б. Аэродинамические производные летательного аппарата и крыла при дозвуковых скоростях, Наука, Москва; 1975. 424 с

[44] Thomas HHBM. Some thoughts on mathematical models for flight dynamics. Aeronautical Journal. 1984;May:169-179

[45] Etkin B. Dynamics of Atmospheric Flight. Mineola, New York: Dover Publications; 2000. 582 p

[46] McRuer D, Ashkenas I, Graham D. Aircraft Dynamics and Automatic Control. Princeton University Press; July 1990

[47] Polhamus EE. A Concept of the Vortex Lift of Sharp-Edge Delta Wings Based on Leading-Edge-Suction Analogy. Washington: NASA; 1966. p. 18

[48] Napolitano M. Aircraft Dynamics: From Modeling to Simulation. J. Wiley & Sons; 2011

[49] Tobak M. On the use of the indicial function concept in the analysis of unsteady motion of wings and wing-tail combinations. NACA Report 1188. 1954. 47 p

[50] Leishman JG. Validation of approximate indicial aerodynamic functions for two-dimensional subsonic flow. Journal of Aircraft. 1988;**25**(10):914-922. DOI: https://doi.org/10.2514/3.45680

[51] Goman MG, Stolyarov GI, Tyrtyshnikov SL, Usoltsev SP, Khrabrov AN. Mathematical description of aircraft longitudinal aerodynamic characteristics at high angles of attack accounting for dynamic effects of separated flow. TsAGI Preprint No. 9; 1990 (in Russian)

[52] Goman M, Khrabrov A. State-space representation of aerodynamic characteristics of an aircraft at high angles of attack. AIAA Paper 92–4651-CP; 1992

[53] Zhang M, Yao Y, Wang H. Aircraft Takeoff Taxiing Model Based on Lagrange Interpolation Algorithm in Theory, Methodology, Tools and Applications for Modeling and Simulation of Complex Systems, Part 2. In: Zhang L, Song X, Wu Y, editors. 16th Asia Simulation Conference and SCS Autumn Simulation Multi-Conference AsiaSim/SCS AutumnSim 2016; Beijing; Oct. 8–11, 2016 Proceedings; Singapore: Springer. 2016. pp. 100-108

[54] Popov AV, Botez RM, Labib M. Transition point detection from the surface pressure distribution for controller design. Journal of Aircraft. 2008;**45**(1):23-28

[55] Dinu AD, Botez RM, Cotoi I. Chebyshev polynomials for unsteady aerodynamic calculations in aeroservoelasticity. Journal of Aircraft. 2006;**43**(1):165-171

[56] Wagner H. Über die Entstehung des dynamischen Auftriebes von Tragflügeln. Zeitschrift für Angewandte Mathematic und Mechanik. 1925;**5**(1):17-35

[57] Theodorsen T. General theory of aerodynamic instability and the mechanism of flutter. Technical Report, Report 496. National Advisory Committee for Aeronautics; 1935

[58] Brunton SL, Rowley CW. Empirical state-space representations for Theodorsen's lift model. Journal of Fluids and Structures. April 2013:174-186

[59] Kurniawan R. Numerical Study of Flutter of a Two-Dimensional Aeroelastic System. Hindawi Publishing Corporation ISRN Mechanical Engineering Volume. 2013. 4 pp. [Article ID 127123]. DOI: http://dx.doi.org/10.1155/2013/127123

[60] Herman JF, Washington ES. Wind Tunnel Investigation of the Aerodynamic Hysteresis Phenomenon on the F-4 Aircraft and Its Effect on Aircraft Motion. US Air Force Report, AEDC-TR-80-10. Aernold Engineering Development Centre. 1980. 95 p

[61] Tobak M, Chapman GT. Nonlinear problems in flight dynamics involving aerodynamic bifurcations. NASA TM 86706; 1985

[62] Yang Z, Igarashi H, Martin M, Hu H. An experimental investigation on aerodynamic hysteresis of a low-Reynolds number airfoil. In: 46th AIAA Aerospace Sciences Meeting and Exhibit; Jan 7–10 2008; Reno, Nevada. AIAA-2008-0315, 11 p

[63] Khrabrov AN, Kolinko KA, Vinogradov YA, Zhuk AN, Grishin II, Ignatyev DI. Experimental investigation and mathematical simulation of unsteady aerodynamic characteristics of a transonic cruiser model at small velocities in a wide range of angles of attack. Visualization of Mechanical Processes: An International Online Journal. 2011;**1**(2). DOI: 10.1615/VisMechProc.v1.i2.40

[64] Ericsson LE. Dynamic stall at high frequency and large amplitude. Journal of Aircraft. 1980;**8**:136-142

[65] Wang Z, Li J. Lan CE, Brandon JM. Estimation of unsteady aerodynamic models from flight test data. AIAA Atmospheric Flight Mechanics Conference and Exhibit, Guidance, Navigation, and Control and Co-located Conferences; Montreal. 2001. https://doi.org/10.2514/6.2001-4017

[66] Harper PW, Flanigan RE. The effect of rate of change of angle of attack on the maximum lift of a small model. Technical Note. NACA TN-2061. 1950. 21 p

[67] Favier D, Agnes A, Barbi C, Maresca C. Combined translation/pitch motion: A new airfoil dynamic stall simulation. Journal of Aircraft. September, 1988:805-813

[68] F/A-18 Stability and Control Data Report. Volume I: Low Angle of Attack. Report # MDC A7247. Issue date 31 August 1981. Revision date 15 November 1982. Revision letter B, McDonell Aircraft Company 3

[69] F/A-18 Stability and Control Data Report. Volume II: High Angle of Attack. Report # MDC A 7247. Issue date 31 August 1981. McDonell Aircraft Company

[70] Katz J, Maskew B. Unsteady low-speed aerodynamic model for complete aircraft configuration. Journal of Aircraft. 1988;April:302-310

[71] Barth TJ, Planeaux P. High angle of attack dynamic behaviour of a model of high performance fighter aircraft. AIAA Paper 88-4368. 1988

[72] Carroll JV, Mehra SB. Bifurcation analysis of nonlinear aircraft dynamics. Journal of Guidance. 1982;**5**:529-536

[73] Mehra SB, Carroll JV. Global stability and control analysis of aircraft at high angles—of—attack. Annual Technical Report, N00014-76-C-0780; Aug. 1979. 356 p. file:///C:/Users/joe/Downloads/ADA084938.pdf, [downloaded at 12 May 2017]

[74] Gao H, He ZD, Zhou ZQ. The study of global stability and sensitivity analysis at high performance aircraft at high angles - of – attack. In: Congress of the International Council of Aeronautical Sciences (ICAS/AIAA); 1988; Jerusalem. 1988. pp. 1356-1363

[75] Jahnke CC, Culick FEC. Application of bifurcation theory to the high-angle-of-attack dynamics of the F-14. Journal of Aircraft. 1994;**31**(1):26-33

[76] Guicheteau P. Bifurcation theory in flight dynamics an application to a real combat aircraft. In: Proceedings of 17th ICAS (International Council of Aeronautical Science) Congress, ICAS-90-5.10.4; Stockholm, Sweden. 1990. pp. 1990-1998

[77] Moes TR, Noffz GK, Iliff KW. Results from F-18B stability and control parameter estimation flight tests at high dynamic pressures. NASA/TP-2000-209033. NASA; 143 p

[78] Cao J, Garret F Jr, Hoffman E, Stalford H. Analytical aerodynamic model of a high alpha research vehicle wind-tunnel model. NASA CR-187469. 1990

[79] Linse DJ, Stengel RF. Identification of aerodynamic coefficients using computational neural networks. Journal of Guidance, Control, and Dynamics (ISSN 0731–5090). 1993;**16**(6):1018-1025

[80] Rajkumar T, Bardina J. Prediction of aerodynamic coefficients using neural network for sparse data. In: Proc. of FLAIRS (Florida Artificial Intelligence Research Society Conference); Pensacola, Florida, US. 2002. pp. 242-246

[81] Secco NR, de Mattos BS. Artificial neural networks to predict aerodynamic coefficients of transport airplanes. Aircraft Engineering and Aerospace Technology. 2017;**89**(2):211-230. DOI: https://doi.org/10.1108/AEAT-05-2014-0069

[82] Ignatyev DI, Khrabov AN. Application of neural networks in simulation of dynamic effects of canard aircraft aerodynamics. TsAGI Science Journal. 2011;**42**:817-828

[83] Rajkumar T, Aragon C, Bardina J, Britten R. Prediction of aerodynamic coefficients for wind tunnel data using a genetic algorithm optimized neural network. In: Zanasi A,

Brebbia CA, Ebecken NFFE, Melli P, editors. Third International Conference on Data Mining, 2002, Bologna, Italy in "Management Information Systems". pp. 473-487

[84] de Visser CC, Mulder JA, Chu QP. Global aerodynamic modeling with multivariate splines, AIAA-2008-7500. In: AIAA Modeling and Simulation Technologies Conference and Exhibit; Honolulu, HI; August 2008

[85] Kouba G, Botez RM, Boely N. Fuzzy logic method use in F/A-18 aircraft model identification. Journal of Aircraft. 2010;**47**(1):10-17. DOI: https://doi.org/10.2514/1.40714

[86] Wang Z, Lan C, Brandon J. Fuzzy logic modeling of nonlinear unsteady aerodynamics. In: 23rd Atmospheric Flight Mechanics Conference, Guidance, Navigation, and Control and Co-located Conferences, AIAA-98-4351; Boston. 1998

[87] Mohammadi SJ, Sabzeparvar M, Karrari M. Aircraft stability and control model using wavelet transforms. Institution of Mechanical Engineers, Part G: Journal of Aerospace Engineering. 2010;**224**:1107-1118

[88] Mader CA, Joaquim R, Martins RA. Computing stability derivatives and their gradients for aerodynamic shape optimization. AIAA Journal. 2014;**52**(11):2533-2546

[89] Brandon JM, Morelli EA. Real-time onboard global nonlinear aerodynamic modeling from flight data. Journal of Aircraft. September;**53**(5):1261–1297

[90] Mavris DN, Kirby MR. Technology identification, evaluation and selection for commercial transport aircraft. Presented at 58th Annual Conference of Society of Allied Weight Engineers, Inc.; San Jose, CA; 24–26 May 1999; Georgia Institute of Technology. SAWE Paper No. 99-2456

[91] Kirby MR. A methodology for technology identification, evaluation and selection in conceptual and preliminary aircraft design [PhD thesis]. Georgia Institute of Technology; 2001

[92] Veress A, Bicsak G. New adaption of actuator disc method for aircraft propeller CFD analyses. Acta Polytechnica Hungarica [Journal of Applied Sciences Hungary]. 2017

[93] Veress Á, Bicsák G, Rohács D. Pressure loss and flow uniformity analysis of baseline and redesigned engine inlet duct for a small turboprop aircraft. Czech Aerospace Proceedings. 2016;**1**:3-9

[94] Rohacs J, Veress A, Jankovics I, Volescsuk A, Farkas Cs. Goal oriented aerodynamic design of a new acrobatic aircraft, research and education in aircraft design. In: READ 2010 International Conference; Warsaw, Poland; June 28-30, 2010; Proceeding CD. 10 p. ISSN 1425-2104

[95] Jankovics IR, Rohacs D, Rohacs J. Motion simulation model of a special acrobatic aircraft. In: Proceedings of the 12th MINI Conference on Vehicle System Dynamics, Identification and Anomalies; Budapest University of Technology and Economics; 2012. pp. 393-401 ISBN: 978-963-313-058-2

[96] Hargitai LC, Rohács D. Motion simulation of inland vessels. Brodogradnja. 2017;**67**

[97] Rohacs D, Rohacs J. Magnetic levitation assisted aircraft take-off and landing (feasibility study—GABRIEL concept). Progress in Aerospace Sciences. 2016;**85**:33-50. DOI: 10.1016/j.paerosci.2016.06.001

[98] Rohacs D, Voskuijl M, Siepenkotter N. Evaluation of landing characteristics achieved by simulations and flight tests on a small-scaled model related to magnetically levitated advanced take-off and landing operations. Proceedings of the 29th Congress of the International Council of the Aeronautical Sciences (ICAS); St. Petersburg, Russia; 07/09/2014–12/09/2014. Paper Rohacs_et al. 9 p

[99] Rohacs D, Voskuijl M, Pool D, Siepenkötter N, Sibilski K. Aircraft magnetic levitated advanced take-off and landing concept validation. Periodica Polytechnica. 2018

[100] Rohács J, Thomasson P, Mosekilde E, Gránásy P. Effects of non-linearities on aircraft poststall motion. In: Nonlinear Problems in Aviation and Aerospace; INCPAA Proceedings; 1996 May, 9–11; Daytona Beach, Florida, USA. 1996. pp. 601–609

[101] Rohács J, Gránásy P. In: Sivasundaram S, editor. Effects of Non-Linearities in Aerodynamic Coefficient on Aircraft Longitudinal Motion, Non-Linear Problems in Aviation and Aerospace. New York: Gordon and Breach; 2000. pp. 281-296

[102] Rohacs J, Bathory ZS. Analysis of approximation of aircraft stochastic motion by markov models. In: ICAS Congress; Yokohama, Japan; CD-ROM, 2004, ICAS. 2004.10.2.1–4.10.2.10. http://www.icas.org/ICAS_ARCHIVE/ICAS2004/PAPERS/324.PDF

[103] Rohács J. Unconventional flight analysis. In: "21st Congress of the International Council of Aeronautical Sciences"; 13–18 September 1998; Melbourne, Victoria, Australia; ICAS Technical Proceedings on CD-ROM, Sept., 1998. A98-31457, Paper 98-1,4,3

[104] Rohács J. Bifurcation analysis of aircraft poststall motion. In: Proceedings of the 7th Mini Conference on Vehicle System Dynamics, Identification and Anomalies, Budapest 2000; Budapest: Department of Railway Vehicles, BUTE. 2001. pp. 53-76

Chapter 10

The Effects of Storage on Turbine Engine Fuels

David W. Johnson

Additional information is available at the end of the chapter

http://dx.doi.org/10.5772/intechopen.69897

Abstract

Modern aviation requires reliable and safe sources of fuel which means fuel is frequently stored for extended periods. In addition, as fuel is used, new fuel is added which is not always compatible with the fuel in the tank. The incompatibility and long-term storage leads to a number of problems that will be addressed in this chapter. Some of the possible changes over time include formation of biofilms, deposit formation, water incorporation and additive depletion. The chemistry and biochemistry of each of these areas will be discussed along with how they might be prevented. New areas of research on low temperature oxidation of trace fuel components and the prevention of bacterial growth will be presented. Other problems are related to the reaction of trace components in the fuel which can lead to oxidation and deposit formation. Trace components also vary based on fuel source and lead to problems in compatibility of different fuels. In addition some of the reactions of fuel additives will be discussed.

Keywords: fuel additive depletion, oxidation, biofilm formation, polar component, deposits

1. Introduction

The needs of modern aviation for a safe and reliable source of fuel are without question. Safe fuels must among other things, be able to be stored for periods of time and be compatible with similar fuels from other sources. For commercial uses, long-term storage is not normally required, since most fuel is used within a few months of refining although longer periods may be required at smaller airports. The military, however maintains much larger stockpiles of fuel in reserve for use in case of emergencies.

The composition of jet fuel can vary significantly depending on the refining process and the source of crude oil. Compatibility can be an issue, since storage tanks are normally refilled when they reach a certain level, and fuels from different suppliers are mixed, even though compatibility issues are possible. Other aspects of storage standards require that the fuel remain free of contaminants, including surfactants, other petroleum products, microorganisms and dyes. All of these types of contaminants can arise from many sources, but careful storage can minimize the hazards associated with some.

The storage and delivery of aviation fuels is carefully regulated and must meet certain standards Including maintaining the purity of the fuel and safety of the aircraft and ground crew [1]. In this chapter, the contamination of fuel from only a few sources will be considered; in particular water contamination, contamination with solids formed by reaction to the fuel with oxygen and contamination due to microorganisms. Finally, some of the challenges associated with the transition to alternate fuels will be discussed.

2. Properties of jet fuel

Turbine engine fuel for aviation uses is a petroleum based fuel similar in properties to kerosene, although newer biofuels and synthetic fuels are under development. Liquid hydrocarbon based fuels are considered ideal for transportation use because they are more easily handled than solids and have a higher energy density than gasses. In most cases, fuel for aviation purposes is of a higher quality than fuels for heating or road transportation purposes. The higher quality requires a greater degree of refining to significantly reduce the level of sulfur, nitrogen and oxygen containing components. It is still thought that the heteroatom containing compounds are responsible for many of the issues of deposit formation and low temperature oxidation.

Fuel for turbine engines is produced meeting one of several specifications. Early turbine engines ran on kerosene based fuel. Over time, a wide cut fuel incorporating gasoline and kerosene was developed for use by the US military, because it was thought availability would be better. It was found, however to increase casualties and evaporation of fuel. It is still currently used in cold regions under the Jet B specification. In the rest of the world, commercial aviation has used a kerosene based fuel under the specification Jet A, Jet A-1 [2] or TS-1. Some

Property	Jet A	Jet A-1	Jet B
Place of use	United States	Western Europe	Parts of Canada and Alaska
Type of fuel	Kerosene	Kerosene	Wide cut
Density	0.775–0.840 g/cm^3	0.775–0.840 g/cm^3	0.750–0.801 g/cm^3
Flash point	38°C	38°C	28°C
Freezing point	−40°C	−47°C	−51°C
Energy content	42.8 MJ/kg	42.8 MJ/kg	42.8 MJ/kg
Aromatics	25%	25%	25%

Table 1. Physical properties of some common jet fuels.

properties of each of the commercial jet fuels are shown in **Table 1**. The US military also has a number of different fuel designations; although JP-5 and JP-8 are the most commonly used and are kerosene based fuels.

While these properties can be met by a number of different fuel sources, the composition of the fuels can be quite different. In many cases, it is the trace heteroatom containing compounds that negatively affect the chemistry of the fuel, especially under storage conditions. In many instances, additives are included in the fuel in order to reduce the reactivity of the heteroatom containing compounds, but the additives themselves may contribute to the reactivity of the fuel. Several of the common additives found in jet fuels are described in **Figure 1** and **Table 2** [3].

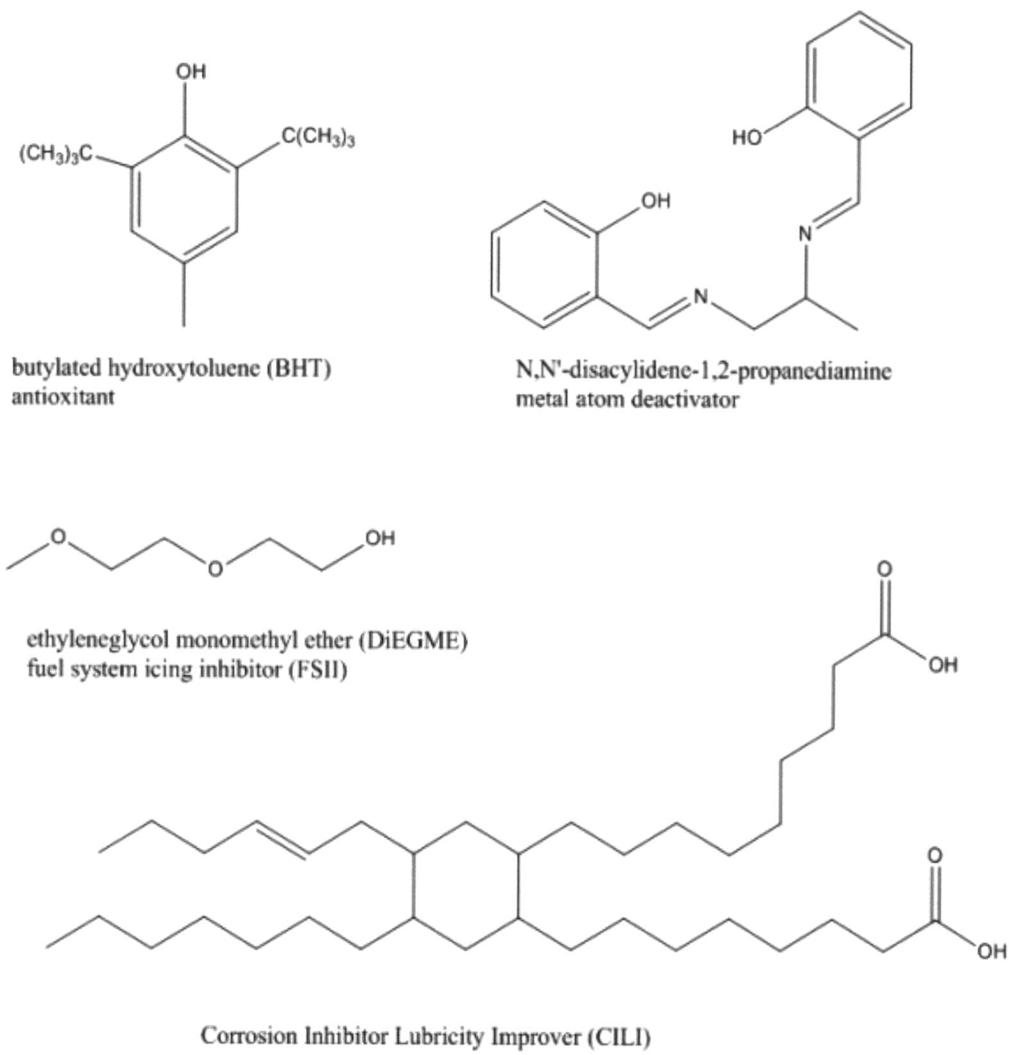

Figure 1. Chemical structures of some common additives found in fuels.

Additive type	Jet A	Jet A-1	JP-4	JP-5	JP-8
Antioxidant	Allowed	Required	Required	Required	Required
Metal deactivator	Allowed	Allowed	Agreement	Agreement	Agreement
Corrosion inhibitor/lubricity improver (CILI)	Agreement	Allowed	Required	Required	Required
Fuel system icing inhibitor (FSII)	Agreement	Agreement	Required	Required	Required

Table 2. Some common additives found in jet fuel.

3. Incorporation of water

3.1. Sources of water in fuels

Water is a very common contaminant in all fuels. The most common ways in which water is incorporated into fuel is through contact with air, condensation of water in the air on cold tank surfaces, leaks in floating cover storage tanks and hatches left open. In most cases, liquid water will drip through the fuel with a small amount dissolving in the fuel. Simple contact with air will cause surface layers of the fuel incorporate water based on the solubility of the water in the fuel, the relative humidity of the air and the temperature. Fuel typically arrives from the refinery saturated with water, which makes reducing the dissolved water in fuel impossible to eliminate. Through mixing, water reaches the entire storage tank. Since the temperature and humidity of the air is constantly changing, the amount of water in the fuel never reaches a steady state.

Kerosene based fuels will dissolve between 40 and 80 ppm water at 20°C. The solubility rises with increasing temperature and varies considerably with the composition of the fuel. Fuels with a high aromatic content can dissolve much more water than those with little aromatic content [4]. Single ringed aromatic compounds, for example dissolve 5–10 times more water than similar saturated compounds. Mixing fuels changes the solubility of water, which can lead to water depositing out of the fuel after mixing.

In addition to dissolved water in jet fuel, there is frequently also free water and emulsified water. Free water is water that sinks to the bottom of the storage tank. Since water is significantly more dense than fuel, if through changes in temperature or fuel composition the fuel exceeds its saturation point water droplets will form and sink to the bottom of the storage tank. As the seasons change is repeated, it is possible for a significant amount of water to settle to the bottom of the tank.

An emulsion is formed when one liquid forms tiny droplets of less than 100 microns that are suspended in another liquid. Many emulsions will spontaneously separate if given enough time, but some will remain suspended indefinitely. Emulsions can form with changes in

http://dx.doi.org/10.5772/intechopen.69897

temperature which can lower the solubility below the saturation point. The can also form when new fuel is added rapidly, dispersing free water into the bulk of the fuel as tiny droplets. Naturally occurring surfactants and surfactants formed during the refining process are known to stabilize water emulsions in jet fuel. These surfactants include naphthenic acids and sulfonic acids and they can cause an emulsion to be stable indefinitely [5].

3.2. Effects of water on fuels

Water can be present in fuel as either dissolved water, emulsified water or free water. Free water is water that is collected at the bottom of the tanks. In a well-designed storage system, the free water collects in well-defined places where it can be periodically drained. The dominant problem with free water that remains in the tank is it forms an interface for microbiological growth, which will be discussed in Section 5. Emulsified water, while it may eventually settle is a more difficult problem since it cannot be conveniently drained and can clog filters and forms a nucleus for ice formation. Typically, emulsified water and free water in the tanks of aircraft are picked up with the fuel and fed to the engine [6].

Dissolved water can be a significant problem since there is no physical way to remove it from the fuel. It is simply pumped into the aircraft. In flight aircraft fuels can experience a wide range of temperatures which certainly change the solubility of water in the fuel. As the temperature of the fuel decreases due to the low temperatures observed at high altitude small droplets of water form. Depending on conditions at low temperature, ice crystals can form and remain suspended in the fuel [7]. In order to reduce the possibility of ice formation fuel system icing inhibitors (FSIIs) have been developed and are required in military jet fuels. Diethylene glycol monomethyl ether (DiEGME) and triethylene glycol monomethyl ether (TriEGME) are two common FSII additives that act by stabilizing water in the fuel, reducing the formation of droplets, which decreases the temperature at which ice can form. A second approach under development is the use of a water reactive compound as an additive to completely remove the water from the fuel system. These compounds are either ketals or ortho esters which spontaneously react with water to form alcohols and either ketones or esters. The products of the reaction with water also act as icing inhibitors [8]. The mechanism of one such additive is shown in **Figure 2** below.

The effect of water on the usability of jet fuel depends greatly on the form of the water. Free water tends to settle to the bottom of the tank in pools. Ribbed tanks are frequently used so that the water will collect in predetermined areas and can be drained from the tank. A problem that can arise from the presence of free water is additive depletion. Some additives such as DiEGME and to a lesser extent TriEGME are much more soluble in water than they are in the fuel and will move from the fuel into the free water resulting in significant or complete additive depletion. Typical fuel systems use a filter or separator to ensure that free water is not pumped into the tank of an aircraft. There is still the possibility of free water depositing from the fuel due to changes in temperature and also changes in fuel source.

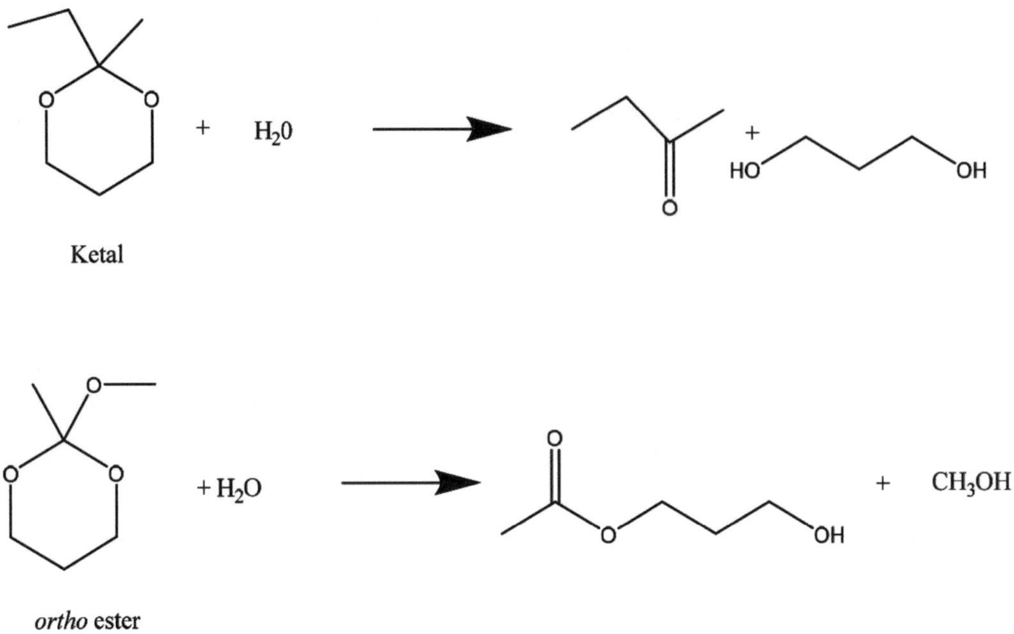

Figure 2. Reaction of a ketal and an orthoester with water.

4. Formation of chemical deposits

Chemical changes in the composition of fuels upon storage are a significant source of contamination and can result in the formation of particulates, slimes and other types of deposits in the fuel. The deposits can clog filters at the fueling stations, or if they pass through the filters cause operational problems in the aircraft. The changes in composition can be the result of the variable composition of fuels, low temperature oxidation of fuel components or other possible reactions of various fuel components. The reaction products are frequently insoluble in the fuel, resulting in their separation as a solid of film. The different types of reactions will be discussed separately in the sections below.

4.1. Low temperature oxidation of fuel components

The oxidation chemistry of jet fuel is normally divided into two distinct regions, with different mechanisms. The auto oxidation mechanism becomes important at about 140°C and the pyrolytic mechanism becomes important at about 300°C. Neither of these regions would appear to be operating at the normal storage temperatures of fuels. Studies of these mechanisms however focused on the chemistry of the hydrocarbon components of the fuel. They indicated that aromatic and cyclic hydrocarbons were substantially more reactive than the paraffinic compounds in the fuels [9]. It is known that free radical autoxidation of hydrocarbons occurs readily at temperatures between 30 and 60°C. The process is thought to be a radical chain reaction initiated by peroxides which abstract a hydrogen atom forming a free radical [10]. It has been shown that a number of factors are involved, but it is often the termination reaction in the radical chain mechanism that leads to the formation of deposits. A startling observation was that the less stable fuels formed less deposit [11]. This observation is likely due to the less stable fuels reacting more quickly, but not through a mechanism that leads to the formation of solid deposits. A series of steps, after the formation of the radical; leads to the formation of oxygenated products, such as primary alcohols [12]. The alcohols further react to form more complex oxygenated products such as dihydrofuranones [13].

What is frequently not accounted for in studies of the autoxidation mechanism of fuels is the presence of trace polar components in the fuel and the presence of catalytic metal ions in the fuel. Polar components have been shown to be correlated to deposit formation, with phenols, indoles and carbazoles showing the largest effect [14]. Some common classes of polar compounds are shown in **Figure 3**. It is well known that nitrogen containing compounds are oxidized much more rapidly that typical hydrocarbons. The enhanced reactivity of the polar compounds then leads to a wide range of new reactions that eventually result in the formation of particulates and films.

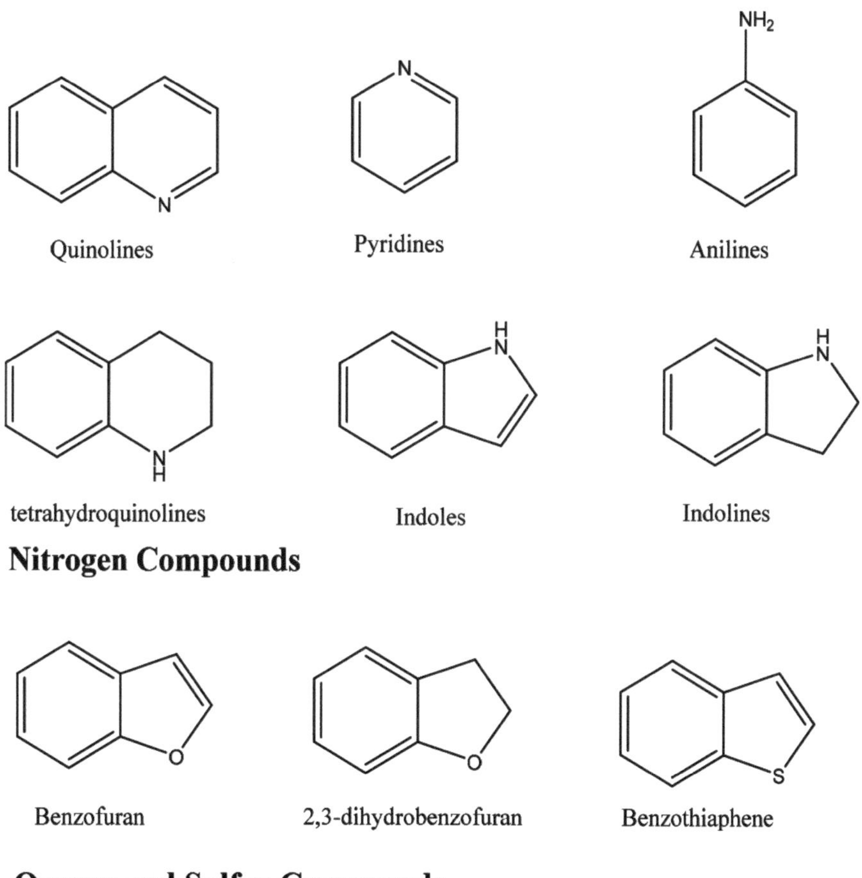

Figure 3. Classes of polar compounds found in jet fuel.

The oxidative addition reactions of a number of nitrogen, oxygen and sulfur containing compounds have been studied at 130°C. Considering that these compounds react completely in a few hours at this temperature, the mechanism could be valid at lower temperatures over a longer period of time. The oxidative addition reaction of several nitrogen and oxygen containing polar compounds are shown in **Figure 4**. It has been shown that the rates of oxidative addition depend greatly on the hetero atom present with N > O >> S [15].

The observation that the oxidative addition reaction produces molecules of significantly higher molecular weight and that those molecules can react further suggests that the problems would eventually become insoluble in the jet fuels, resulting in the formation of solid particles or oligomeric films.

A fundamental understanding of the mechanism of deposit formation under both thermal autoxidation and typical storage conditions has been a long-term goal. One possible mechanism that leads to the formation of high molecular weight products involves the oxidation of hydrogen containing heteroatomic aromatic compounds to form electrophilic quinone-like species. The quinone-like species reacts with nucleophile present in the fuels to eventually form soluble macromolecular oxidatively reactive species (SMORS) [16]. The initial step of the SMORS mechanism is the reaction of phenol with a peroxy radical to abstract a hydrogen atom. A similar reaction would also be possible with anilines and thiophenols that may

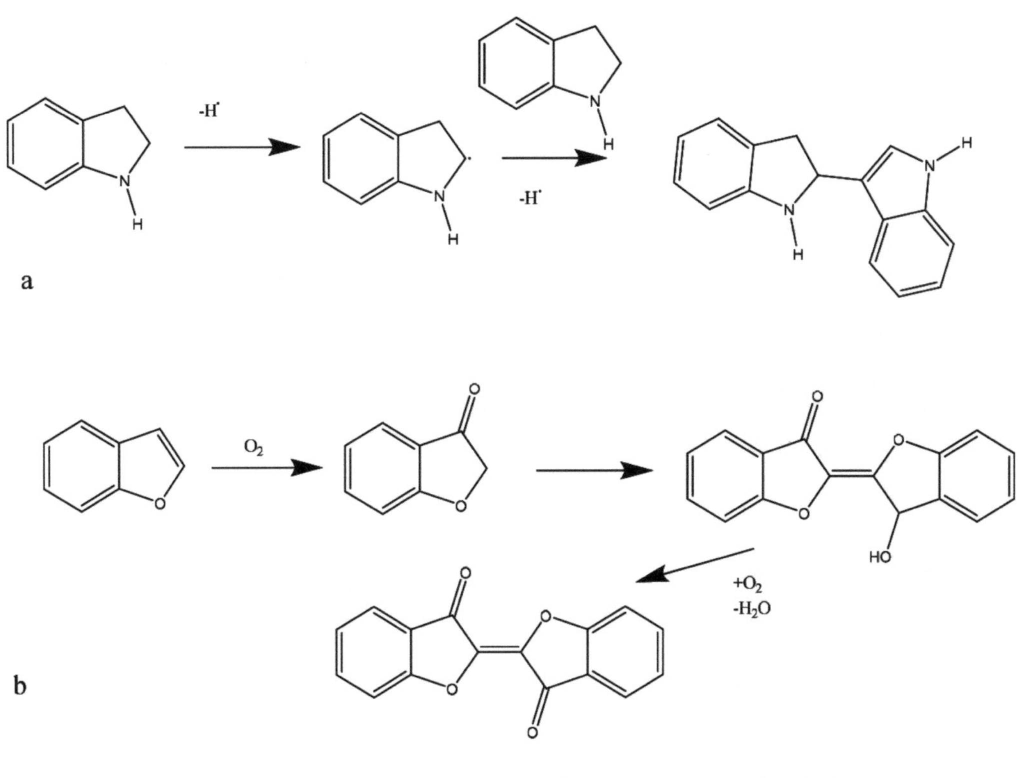

Figure 4. Reactions of polar compounds in jet fuel. (a) Reaction of 2, 3-dehydroindole to form higher molecular weight products and (b) reaction of benzofuran to form higher molecular weight products.

also be present [17]. The importance of the phenol has been demonstrated by conversion of active hydrogen containing species to silyl ethers which resulted in a dramatic decrease in deposit formation [18]. The phenoxy radical rapidly reacts to form an electrophilic quinone which then reacts with a nucleophilic aromatic heterocyclic compounds, such as pyroles, indoles, and carbazoles (see **Figure 5**). Further reactions lead to the formation of SMORS which can then react further to form insoluble deposits [19]. The understanding of this mechanism has led to some possible stabilizers which may improve the stability of stored fuel [20].

4.2. Reactions between fuels with differing composition

Jet fuels obtained from different sources, refined in different ways have been shown to have somewhat different compositions. The differences are quite apparent when the classes of polar compounds from several different jet fuel samples are compared as is shown in **Table 3**.

The heteroatom containing components of the fuels are in large part responsible for the low temperature reactivity of the fuel [21]. This can be particularly problematic when a fuel that is high in basic nitrogen containing groups is added to a storage tank containing a fuel rich in acids or phenols. The product of the acid-base reaction is reasonably expected to form an insoluble film in the tank.

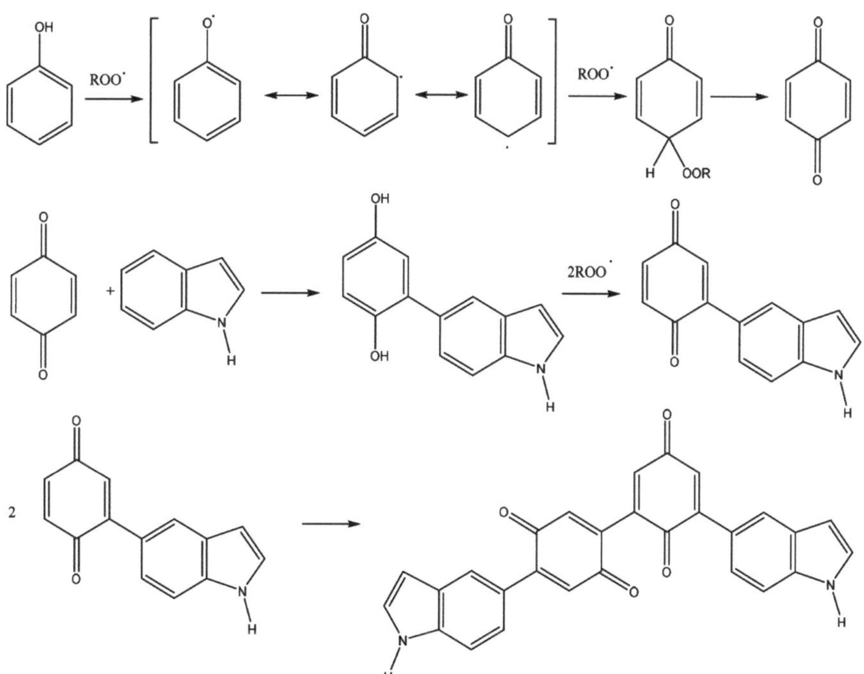

Figure 5. Mechanism for the formation of SMORS, and further reaction to form insoluble species. The figure shows the reaction of indole, however other nucleophilic heterocyclic compounds also react.

Fuel	Type	Phenols	Quinoline/ aminonaphthalene	Indoline/ tetrahydroquinoline	Aniline/ pyridine	Pyrroles/satd. indoles
2747	Jet A	X	X	X	X	
2959	Jet A	X				
3656	Jet A	X	X	X	X	X
3658	Jet A	X	X	X	X	X
3773	JP-8	X	X	X	X	
4336	JP-8	X	X	X	X	X
4177	JP-8	X	X	X	X	
5098	JP-8	X	X	X	X	X

Table 3. Comparison of the classes of polar compounds found in various jet fuels.

5. Biological contamination and biofilm formation

Microbial contamination of fuels is a major problem that affects all types of fuels, however because of the demanding specifications and use requirements for jet fuel is a much more important type of contamination. Jet fuel is a very harsh environment; however, many different types of microorganisms have found ways to metabolized different components of the fuel. The paraffinic and some other hydrocarbons are a food source for bacteria but some aromatic compounds and most sulfur containing compounds are toxic to microorganisms [22]. Microorganisms are generally found at the water-fuel interface, indicating that water removal is always a part of microorganism control. The effects of microorganisms growing in fuel are quite diverse, from the filter clogs in the fueling system, to problems in flight, to biocorrosion problems in the tanks and associated fueling system components and hydrogen sulfide corrosion [23].

Microorganisms that can metabolize a wide range of aromatic and aliphatic hydrocarbons have been identified and studied. The metabolism of aromatic compounds involves initial attack on side chains to oxidize the compound at the benzylic position and in subsequent steps open the aromatic ring [24]. The metabolism of paraffinic hydrocarbon is known to occur under both aerobic and anaerobic conditions and there have been several mechanisms proposed [25]. All of the mechanisms appear to convert the hydrocarbons initially to fatty acids either through an oxidative mechanism or through a mechanism that begins with the addition of fumarate, followed by a series of steps to give carbon dioxide and fatty acids [26]. The various anaerobic bacteria depend upon either sulfate or nitrate as an oxidant in their metabolism.

In order to survive, microorganisms have mechanisms that protect them from the toxic compounds such as formation of biofilms, secretion of surfactants and regulation of efflux pumps [27]. Biological contamination, especially where colonies or films can clog filters, reduce fuel flow among other effects. Biological contamination also changes the composition of the fuel, as the microorganisms metabolize their preferred hydrocarbons, typically paraffinic compounds, leaving the more difficult to metabolize. Some of the effects of microorganisms on fuel and fuel storage systems are shown in **Table 4**.

A number of different types of microorganisms have been isolated from air force fuel tanks, including *Micrococcus, Bacillus, Staphylococcus, Sphingomonas* and *Discosphaerina fagi* [28]. Microorganisms are typically present in fuels, but require the presence of water in order to grow. Regular removal of water and the use of biocides can minimize their growth [29]. Numerous types of microorganisms have been isolated from fuel systems, including moulds, bacteria, yeasts and fungi. The microorganisms require water for growth and they feed on nutrients in the fuel. Many microorganisms can metabolize the hydrocarbons, while others feed on partially degraded fuel, the trace heteroatom containing contaminants in the fuel and the additives in the fuel [30].

The types of microorganisms present in the fuel frequently depend on storage conditions and length of storage. Normally, aerobic microorganisms are dominant; since there is a constant supply of oxygen saturated fuel. During long-term fuel storage however, the oxygen is quickly used and anaerobic microorganisms flourish. In many cases, bacteria are able to use sulfate which results in the formation of hydrogen sulfide. This toxic and foul smelling gas attacks the steel of the bottom plate. Hydrogen sulfide also dissolves in the fuel and the fuel can become aggressive toward steel, silver and copper alloys [31]. Anaerobic bacteria are a particular problem in marine environments where salt water provides an abundant supply of sulfate that the anaerobic bacteria convert to hydrogen sulfide which leads to foul smelling and toxic fuel and significant biocorrosion of steel [32].

Problem	Microorganism type
Blockage of pipes, valves, filters and incorrect readings from fuel probes	Fungi; polymer producing bacteria
Increased water content	All
Sludge formation	All
Surfactant production, forming fuel/water emulsions and coalesce failure	Fungi and aerobic bacteria
Corrosion of storage tanks	Fungi and anaerobic bacteria
Production of suspended solids	All
Breakdown of hydrocarbons	Fungi and aerobic bacteria
Fouling of injectors	Fungi and aerobic bacteria
Penetration of protective linings	Fungi
Increased sulfur content in fuel	Sulfur-reducing bacteria
Health problems	Endotoxin producing bacteria, pathogens, sulfur reducing bacteria

Table 4. The effects of microorganisms on fuel system components.

Microorganisms have been shown to be highly adaptable in their ability to use different food sources. It has been shown that several adaptations are needed for Pseudomonas aeruginosa to metabolize jet fuel [33]. When jet fuel is the available carbon source, the bacteria alter their metabolism through transcriptional regulation to favor the use of paraffinic hydrocarbons of the C_{11}–C_{13} length as a food source [34]. These same transcriptional changes require biofilm formation for bacterial growth.

The introduction of ultralow sulfur and low sulfur fuels introduces new complexities in biofilm formation. It is known that sulfur compounds normally found in petroleum based fuels are natural lubricity improvers, antioxidants and antimicrobial agents [35]. Studies have shown that the removal of the sulfur compounds does not appear to alter the bio-corrosion properties of the fuel under anaerobic conditions [36], however microorganisms have also been shown to rapidly deplete the corrosion inhibitor/lubricity improver (CILI) additives from the fuel [22].

On approach to reduce microorganism growth in fuels is to incorporate biocides in the fuel formulation. There are very few biocides that have been approved for use in Jet A, but are not allowed in the military jet fuels JP-4, JP-5 and JP-8. In Jet A, the allowed biocides are Biobor™ and Kathon™. Kathon has a sulfur heterocyclic compound as the active ingredient [37] and Biobor has a boron containing compound [38].

6. Alternative or synthetic fuels

The aviation community uses huge amounts of fuel and requires a reliable source of high quality fuel. The uncertain markets for petroleum and the differences in fuel composition due to different crude oil feed stocks has led the industry to investigate alternate sources of fuel. In the United States, several alternative energy sources are possible, including natural gas, coal and biomass. Natural gas and coal are in abundant supply in the United States but are still non-renewable and have a large environmental footprint [39]. Biomass based sources, are potentially more environmentally friendly, although the life cycle analysis must certainly be taken into account [40, 41]. The United States military has estimated that at least 50% of its energy use would be from renewable sources by the year 2020 [42]. Several processes are under investigation, but considering current market conditions of low crude oil prices, prospects for widespread development are not promising [43].

Primary attention has been paid to processes which produce fuels that can serve as drop in replacements for petroleum based fuels requiring no modifications to the aircraft, and can be interchanged with conventional fuel depending upon local availability [44]. It is also important that the feedstock can be produced sustainably and are not either used in foods or raised on land that can also be used to raise food crops [45]. Two processes that are the most developed and can produce a synthetic fuel from a variety of feedstocks to provide an alternative fuel source are synthetic fuels through the Fisher-Tropsch process and hydroprocessed esters and fatty acids fuels.

http://dx.doi.org/10.5772/intechopen.69897

The synthesis of liquid fuels from natural gas, coal or biological sources has been an important goal for many years. The primary industrial process that is used is the Fischer-Tropsch process that was initially developed in Germany in the early 1920s. The process takes carbon monoxide and hydrogen and converts them in the presence of a catalyst to paraffinic and branched chain hydrocarbons. The carbon monoxide and hydrogen can come either from coal, natural gas or renewable sources. Depending on the catalyst and reaction temperature different types of fuel can be prepared [46]. Fischer-Tropsch fuels have the advantage of containing no aromatic content, no metals and no heteroatom containing impurities. These fuels have been shown to be cleaner burning with reduced particulate emission than conventional fuels. Current specifications for aviation use state that synthetic fuels must either be mixed 50–50 by volume with petroleum based fuels for semi-synthetic jet fuel or be tested to ensure at least 8% aromatic content for a fully-synthetic jet fuel.

The aviation industry has shown an intense interest in developing fuels from renewable resources. Initial evaluation of fatty acid methyl esters similar to biodiesel showed the fuel did not have low temperature properties needed for aviation [47]. The need however for a renewable source of jet fuel, preferably from a non-food source led to the development of a hydrotreated renewable jet fuel (HRJ) from camelina. The hydrotreating process for jet fuel converts the typical esters into paraffins and isoparaffins by reaction with hydrogen in the presence of a catalyst. Camelina based HRJ and other isoparaffin-rich bioderived fuels are similar in composition to ultralow sulfur hydrocarbon fuels [48]. Isoparaffin rich fuels also have the advantage of lower exhaust emission of nitrogen oxides and particulate matter.

The storage stability of Fischer Tropsch fuels and hydrotreated renewable jet fuel will be likely enhanced by the reduced solubility of water in fuels that have no aromatic content. Since the heteroatoms are not present, many of the deposit formation reactions found in normal fuels should be absent. The primary concern would be the ability of microorganisms to metabolize the hydrocarbons that comprise the majority of the fuel. Microorganisms have been shown to metabolize synthetic paraffinic kerosene readily, since aromatics and sulfur containing compounds in conventional fuels are toxic to many of the microorganisms. The use of synthetic jet fuel has been limited to a partially synthetic 50:50 blend of the synthetic fuel with a petroleum jet fuel; with the exception of a fully-synthetic fuel produced by Sasol. The primary limitation is due to the lack of aromatics which can result in seal cracking. The added petroleum based jet fuel or additional treating in the SASOL product increases the aromatic content to 8% which is adequate to avoid seal shrinkage [49]. The increased aromatic content, however leads to an increase in oxidative addition reactions which are closely associated with the formation of deposits [50].

7. Conclusions

Civilian and military aviation requires an abundant, affordable, safe and storable source of fuel. The variable composition of fuel depending upon crude oil source and refining methods creates many challenges for fuel storage. Chemical deposit formation, which has been tied to oxidation reactions, followed by addition mechanisms is one source of fuel degradation. Other problems in fuel storage are related to the presence of water, both from contamination through leaks in the storage tanks and the general solubility of water in jet fuel can create icing problems and also provide a growth medium for microorganisms. Good water management can provide added safety in the fuel storage system. Finally, the development of renewable and synthetic fuels will change the problems of fuel storage, due to the more predictable composition of the fuel and the lack of heteroatom containing impurities associated with the fuels.

Author details

David W. Johnson

Address all correspondence to: djohnson1@udayton.edu

Department of Chemistry, University of Dayton, Dayton, Ohio, United States of America

References

[1] Standard for Jet Fuel Quality Control at Airports, Revision 2016, ATA Specification 103. Washington, D.C.: Air Transport Association of America; 2016. Available from: http://infostore.saiglobal.com/store/details.aspx?ProductID=1858071

[2] Turbine Fuel, Aviation Kerosene Type, Jet A-1. Defence Standard 91-091, Issue 9. UK: Ministry of Defence; October 2016

[3] The Chevron Products Group. Aviation Fuels: A Technical Review. Chevron, USA; 2007. San Ramone, CA. Available from: https://www.cgabusinessdesk.com/document/aviation_tech_review.pdf

[4] Lam JK-W, Carpenter MD, Williams CA, Hetherington JI. Water solubility characteristics of current aviation jet fuels. Fuel. 2014;**133**:26-33. DOI: 10.1016/j.fuel.2014.04.091

[5] Carpenter MD, Hetherington JI, Lao L, Ramshaw C, Yeung H, Lam JK-W, Masters S, Barley S. Behavior of water in aviation fuels at low temperatures. In: Morris RE, editor. Proceedings of the 12th International Conference on Stability, Handling and Use of Liquid Fuels; October 16-20, 2011; Sarasota, FL, United States. pp. 1-18

[6] Baena-Zambrana S, Repetto SL, Lawson CP, Lam JK-W. Behavior of water in jet fuel – A literature review. Progress in Aerospace Sciences. 2013;**60**:35-44. DOI: 10.1016/paerosci.2012.12.001

[7] Lam JK-W, Hetherington JI, Carpenter MD. Ice growth in aviation jet fuel. Fuel. 2013;**113**:402-406. DOI: 10.1016/j.fuel.05.048

[8] Repetto SL, Patel R, Johnson T, Costello JF, Lam JK-W, Chuck CJ. Dual action additives for Jet A-1: Fuel dehydrating icing inhibitors. Energy & Fuels. 2016;**30**:9080-9088. DOI: 10.1021/acs.energyfuels.6b01727

[9] Zabarnick S, Phelps DK. Density functional theory calculations of the energies and kinetics of jet fuel autoxidation reactions. Energy and Fuels. 2006;**20**:488-497. DOI: 10.1021/ef0503481

[10] Mayo FR. Free-radical autoxidation of hydrocarbons. Accounts of Chemical Research. 1968;**1**:193-201. DOI: 10.1021/ar50007a001

[11] Henegan SP, Zabarnick S. Oxidation of jet fuels and the formation of deposits. Fuel. 1994;**73**:39-43. DOI: 10.1016/0016-2361(94)90185-6

[12] Webster RL, Evans DJ, Marriott PJ. Detailed chemical analysis using multidimensional gas chromatography-mass spectrometry and bulk properties of low temperature oxidized jet fuels. Energy and Fuels. 2015;**29**:2059-2066. DOI: 10.1021/acs.energyfuels.5b00264

[13] Webster RL, Evans DJ, Marriott PJ. Multidimensional gas chromatographic analysis of low temperature oxidized jet fuels: Formation of alkyldihydrofuranones. Energy and Fuels. 2015;**29**:2067-2073. DOI: 10.1021/acs.energyfuels.5b00265

[14] Balster LM, Zabarnick S, Striebich RC, Shafer LM, West ZJ. Analysis of polar species in jet fuel and determination of their role in autoxidative deposit formation. Energy and Fuels. 2006;**20**:2564-2571. DOI: 10.1021/ef0602751

[15] Siddiquee MN, de Klerk A. Heterocyclic addition reactions during low temperature autoxidation. Energy and Fuels. 2015;**29**:4236-4244. DOI: 10.1021/acs.energyfuels.5b00767

[16] Beaver B, Gao L, Burgess-Clifford C, Sobkowiak M. On the mechanisms of formation of thermal oxidative deposits in jet fuel. Are unified mechanisms possible for both storage and thermal oxidative deposit formation in middle distillate fuels?. Energy and Fuels. 2005;**19**:1574-1579. DOI: 10.1021/ef040090j

[17] Kabana CG, Botha S, Schmucker C, Woolard C, Beaver B. Oxidative stability of middle distillate fuels. Part 1: Exploring the soluble macromolecular oxidatively reactive species (SMORS) mechanism with jet fuels. Energy and Fuels. 2011;**25**:5145-5157. DOI: 10.1021/ef200964z

[18] Zabarnick S, Mick MS, Striebich RC, Grinstead RR. Model studies of silylation agents as thermal-oxidative jet fuel additives. Energy and Fuels. 1999;**13**:154-159. DOI: 10.1021/ef980149k

[19] Aksoy P, Gul O, Cetiner R, Fonseca DA, Sobkowiak M, Falcone-Miller S, Miller BG, Beaver B. Insight into the mechanism of middle distillate fuel oxidative degradation. Part 2: On the relationship between jet fuel thermal oxidative deposit, soluble macromolecular oxidatively reactive species, and smoke point. Energy and Fuels. 2009;**23**:2047-2051. DOI: 10.1021/ef8007008

[20] Gul O, Cetiner R, Griffith JM, Wang B, Sobkowiak M, Fonseca DA, Aksoy P, Miller BG, Beaver B. Insight into the mechanisms of middle distillate fuel oxidative degradation. Part 3: Hydrocarbon stabilizers to improve jet fuel thermal oxidative stability. Energy and Fuels. 2009;**23**:2052-2055. DOI: 10.1021/ef800701u

[21] Adams RK, Zabarnick S, West ZJ, Striebich RC, Johnson DW. Chemical analysis of jet fuel polar, heteroatomic species via high performance liquid chromatography with electrospray ionization-mass spectrometric detection. Energy and Fuels. 2013;**27**:2390-2398. DOI: 10.1021/ef3015298

[22] Stamper DM, Morris RE, Montgomery MT. Depletion of lubricity improvers from hydrotreated renewable and ultralow-sulfur petroleum diesels by marine microbiota. Energy and Fuels. 2012;**26**:6854-6862. DOI: 10.1021/ef301158n

[23] Roberts GAH. Microbiological corrosion of tanks in long-term storage of gas oil. British Corrosion Journal. 1969;**4**:318-321. DOI: 10.1179/000705969798325000

[24] Heider J, Spormann AM, Beller HR, Widdel F. Anaerobic bacterial metabolism of hydrocarbons. FEMS Microbiology Reviews. 1999;**22**:429-473. DOI: 10.1111/j.1574-6976.1998.tb00381.x

[25] Wentzel A, Ellingsen TE, Kotlar H-K, Zotchev SB, Throne-Huist M. Bacterial metabolism of long-chain n-alkanes. Applied Microbiology and Biotechnology. 2007;**76**:1209-1221. DOI: 10.1007/s00253-007-1119-1

[26] Abbasian F, Luckington R, Megharaj M, Naidu R. A review on the genetics of aliphatic and aromatic hydrocarbon degradation. Applied Biochemistry and Biotechnology. 2016;**178**:224-250. DOI: 10.1007/s12010-015-1881-y

[27] Vu B, Chen M, Crawford RJ, Ivanova EP. Bacterial extracellular polysaccharides involved in biofilm formation. Molecules. 2009;**14**:2535-2554. DOI: 10.3390/molecules14072535

[28] Rauch-Johnson ME, Graef HW, Rozenzhak SM, Jones SE, Bleckmann CA, Naik RR, Stone MO. Identification of microorganisms isolated from United States Air Force aviation fuel sump samples. In: Morris RE, Hughes VB, editors. Proceedings of the 8th International Conference on Stability and Handling of Liquid Fuels; September 14-19, 2003; Steamboat Springs, CO, United States; 2003. pp. 772-779

[29] Gaylarde CC, Bento FM, Kelley J. Microbial contamination of stored hydrocarbon fuels and its control. Revista de Microbiologia. 1999;**30**:1-10. DOI: 10.1590/S0001-37141999000100001

[30] Hill T. Microbial growth in aviation fuel. Aircraft Engineering and Aerospace Technology. 2003;**75**:497-502. DOI: 10.1108/00022660310492582

[31] Tripathi RP, Gulati IB, Pandey SS, Rohatgi HS. Copper and silver corrosion by aviation turbine fuels. Industrial & Engineering Chemistry Product Research and Development. 1973;**12**:227-231. DOI: 10.1021/i360047a013

[32] Liang R, Aktas DF, Aydin E, Bonifay V, Sunner J, Suflita JM. Anaerobic biodegradation of alternative fuels and associated biocorrosion of carbon steel in marine environments. Environmental Science and Technology. 2016;**50**:4844-4850. DOI: 10.1021/acs.est.5b06388

[33] Gunasekera TS, Striebich RC, Mueller SS, Strobel EM, Ruiz ON. Transcriptional profiling suggests that multiple metabolic adaptations are required for effective proliferation of *Pseudomonas aeruginosa* in jet fuel. Environmental Science & Technology. 2013;**47**:13449-13458. DOI: 10.1021/es403163k

[34] Balster LM, Mueller SS, Bowen LL, Brown LM, Vangsness MD, Tsao M, Shafer LM. Microbial growth and biofilm formation: GC-MS analysis and traditional culture of jet A and alternative fuels. In: 41st Central Regional Meeting of the American Chemical Society; June 16-19; Dayton, OH, United States. 2010

[35] Heldreth B, Turos E. Microbiological properties and modes of action of organosulfur-based anti-infectives. Current Medicinal Chemistry – Anti-Infective Agents. 2005;**4**:295-315. DOI: 10.2174/1568012205774322250

[36] Lyles CN, Aktas DF, Duncan KE, Callaghan AV, Stevenson BS, Suflita JM. Impact of organosulfur content on diesel fuel stability and implications for carbon steel corrosion. Environmental Science and Technology. 2013;**47**:6052-6062. DOI: 10.1021/es4006702

[37] Biocide Selection Guide for Fuel Applications. Dow Chemical Company. 2015. Midland, MI. Available from: http://msdssearch.dow.com/PublishedLiteratureDOWCOM/dh_0853/0901b8038085332a.pdf?filepath=microbial/pdfs/noreg/253-02864.pdf&fromPage=GetDoc Downloaded January 30, 2017

[38] Biobor JF, Safety Data Sheet. Hammonds Fuel Additives, Inc. 2015. Available from: http://www.biobor.com/Biobor-Resources/msds/SDS-BIOBORJF-GHS_EU-rev-8-15.pdf Downloaded January 30, 2017, Houston, TX

[39] Blakey S, Rye L, Wilson CW. Aviation gas turbine alternative fuels: A review. Proceedings of the Combustion Institute. 2011;**33**:2863-2885. DOI: 10.1016/j.proci.2010.09.011

[40] Marsh G. Biofuels: An aviation alternative. Renewable Energy Focus. 2008;**9**:48-50. DOI: 10.1016/S1471-0846(08)70138-0

[41] Han J, Elgowainy A, Cai H, Wang MQ. Life-cycle analysis of bio-based aviation fuel. Bioresource Technology. 2013;**150**:447-456. DOI: 10.1016/j.biortech.2013.07.153

[42] Blakely K. DOD alternative fuels: Policy, initiatives and legislative activity. Congressional Research Service Report of Congress. December 14, 2012. Available from: https://ia801304.us.archive.org/31/items/202467DODAlternativeFuelsPolicyInitiativesandLegislativeActivity-crs/202467 DOD Alternative Fuels_ Policy, Initiatives and Legislative Activity.pdf [Accessed: January 20, 2017]

[43] Mawhood R, Gazis E, de Jong S, Hoefnagels R, Slade R. Production pathways for renewable jet fuel: A review of commercialization status and future prospects. Biofuels, Bioproducts and Biorefining. 2016;**10**:462-484. DOI: 10.1002/bbb.1644

[44] Maurice LQ, Lander H, Edwards T, Harrison III WE. Advanced aviation fuels: A look ahead via a historical perspective. Fuel. 2001;**80**:747-756. DOI: 10.1016/S0016-2361(00)00142-3

[45] Rye L, Blakey S, Wilson CW. Sustainability of supply or the planet: A review of potential drop-in alternative aviation fuels. Energy and Environmental Science. 2010;**3**:17-27. DOI: 10.1039/b918197k

[46] Schulz H. Short history and present trends of Fischer-Tropsch synthesis. Applied Catalysis A. 1999;**186**:3-12. DOI: 10.1016/S0926-860X(99)00160-X

[47] Bhale PV, Deshpande NV, Thrombe SB. Improving the low temperature properties of biodiesel fuel. Renewable Energy. 2009;**34**:749-800. DOI: 10.1016/j.renene.2008.04.037

[48] Shonnard DR, Williams L, Kalnes TN. Camelina-derived jet fuel and diesel: Sustainable advanced biofuels. Environmental Progress and Sustainable Energy. 2010;**29**:382-392. DOI: 10.1021/ep.10461

[49] Corporan E, Edwards T, Shafer L, DeWitt MJ, Klingshirn C, Zabarnick S, West Z, Striebich R, Graham J, Klein J. Chemical, thermal stability, seal swell and emissions studies of alternative jet fuels. Energy and Fuels. 2011;**25**:955-966. DOI: 10.1021/ef101520v

[50] DeWitt MJ, West Z, Zabarnick S, Shafer L, Striebich R, Higgins A, Edwards T. Effects of aromatics on the thermal-oxidative stability of synthetic paraffinic kerosene. Energy and Fuels. 2014;**28**:3696-3703. DOI: 10.1021/ef500456e

Chapter 11

12-Pulse Active Rectifier for More Electric Aircraft Applications

Mohamad Taha

Additional information is available at the end of the chapter

http://dx.doi.org/10.5772/intechopen.70882

Abstract

The Aircraft industry is moving very quickly towards what it known as More Electric Aircraft (MEA). In a modern aircraft power technology system instead of using a fixed 400 Hz supply, a variable frequency supply (360 to 800 Hz) is used, which is dependent on the aircraft speed. In MEA electrical energy feeds the aircraft subsystems such as the flight control actuation, environmental control system, and utility function instead of mechanical, hydraulic and pneumatic energy. Although the new technology of MEA goes towards variable frequency supply, one of the essential parts of the power distribution systems require DC power sources to feed different DC loads, and a portion of load may require a fixed 400 Hz supply. This chapter simulation model of 12-pulse active rectifier for More Electric Aircraft applications.

Keywords: MEA, THD, active rectifier, boost converter, power factor, vector control

1. Introduction

Within the modern aircraft industry, More Electric Aircraft technology is growing rapidly. **Figure 1** shows a general block diagram for MEA power distributions. The power distribution system model consists of power generation unit, transformer rectifier unit, DC-DC converter unit, and DC-AC inverter unit.

The satisfactory performance of MEA depends to a very great degree on the continuing reliability of electrical systems and subsystems. This technology has many benefits and advantages such as:

a. Decreasing maintenance and operating costs.

b. Increasing dispatch reliability.

c. Reducing gas emissions.

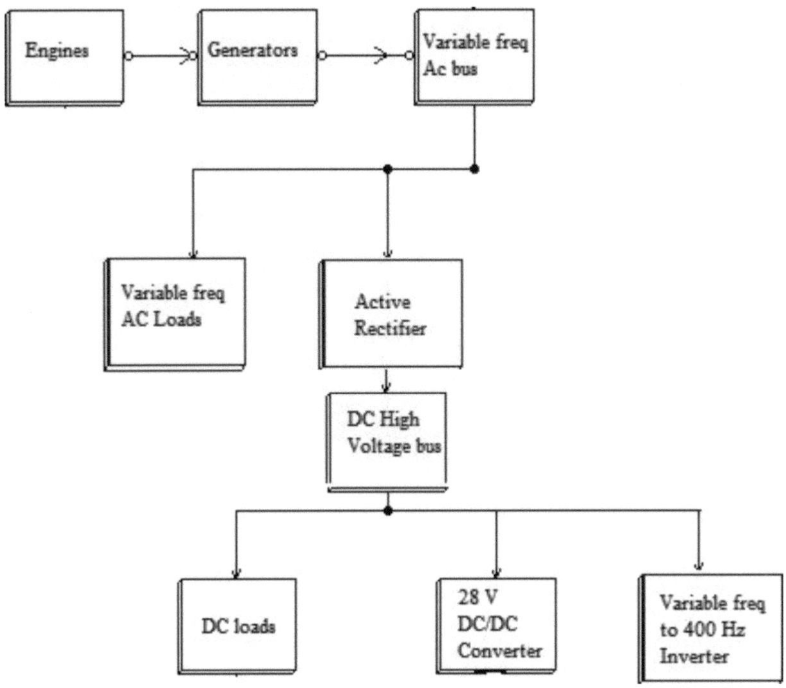

Figure 1. MEA general power distribution system.

In MEA Power Electronics segment plays a very important part in controlling the energy and improving both generators and actuators energy conversion. Furthermore, in a fixed frequency system (400 Hz) a mechanical constant speed drive set between the engine and the aircraft generator, however this will give extra weight and must be frequently maintained.

The use of Power Electronics helps in reducing weight, is easier to maintain, and provides more controllability and intelligence which includes fault detection and diagnosis [1–6].

A conventional 12 pulse rectifier using Diode Bridge is one of the simplest converter since does not require any control loop, however, this type of converter has a fixed DC output with high Total Harmonic Distortion (THD) on the input current compared with the proposed 12-pulse active rectifier.

The system has the ability to stabilize an output voltage of variable V_{dc} from a 3 phase 360–800 Hz, 115 V RMS system. Using a decoupling feed-forward control method by DQ frame technique, the magnitude and the phase of the input current can be controlled and hence the power transfer that occurs between the AC and DC sides can also be controlled. The converter could be suitable to use with an electric actuator (or other) aircraft loads. The system could be used as DC source for DC loads or to feed DC to AC inverter for a fixed 400 Hz supply. The design of this system poses significant challenges due to the nature of the load range and supply frequency variation and requires many features such as:

1. Sinusoidal and low harmonics contents on supply current.

2. High input power factor must be achieved to minimize reactive power requirements.
3. Power density must be maximized for minimum size and weight [7, 8].

Generally, use of electrical power on board is continuously increasing within the areas of communications, surveillance and general systems, such as: radar, cooling, landing gear or actuators systems. DC voltage of up to 540 V [9] may be required for electric power distribution to feed certain loads.

2. Harmonics and power factors for different rectifiers

Non-linear loads such as rectifiers can cause harmonics on the aircraft electrical supply systems, this will increase losses and can excite resonance in some circuits resulting in elevated voltages. For a resistive load, the harmonics current are proportional to the voltage harmonics. For a capacitive load the harmonics current will increase the capacitor heating and can cause premature failure.

For an inductive load harmonics increase losses in core components and also rotor losses in an induction motor will increase. Furthermore harmonics current could overheat transformers, therefore transformers, should be derated in the presence of harmonics.

In general rectifiers produce harmonics with the following order [10–14]:

$$h = \frac{f_h}{f_1} = K.P \pm 1 \tag{1}$$

where h = order of harmonics; f_h = frequency of the harmonic current; f_1 = fundamental frequency; P = rectifier pulse number; K = 1, 2, 3,….

The amplitude of the harmonic currents caused by rectifier can be calculated as:

$$I_h = \frac{I_1}{h} \tag{2}$$

where I_h = amplitude of harmonic current order; I_1 = amplitude of the fundamental current of the rectifier.

In AC power systems with pure sinusoidal voltage and current, the cosine of the phase difference (φ) between the voltage and current represents the power factor (PF = cos φ). If the voltage or current waveforms contain harmonics, the phase angle between them is no longer represents the power factor. In general, the power factor could be calculated as [11].

$$PF = \frac{\text{mean power}}{V_{rms} I_{rms}} \tag{3}$$

Rectifiers draw non-sinusoidal current and have high harmonic components, however, if the input voltage of the rectifier is considered to be a pure sinusoidal, therefore the mean power will be:

$$P_{mean} = V_{rms} I_{1rms} \cos \varphi_1 \tag{4}$$

Therefore:

$$PF = \frac{I_{1rms}}{I_{rms}} \cos \varphi_1 \tag{5}$$

Where $\frac{I_{1rms}}{I_{rms}}$ is defined as the input distortion factor; I_{1rms} is the RMS value of the fundamental current; cos φ_1 is the phase angle between the voltage and the fundamental current (input displacement factor).

Electronic devices in MEA technology are increasing, which are usually powered by switched mode power supplies (SMPS). SMPS will properly feed from a diode rectifier which imposes harmonic currents and possibly voltages onto the mains power network on the aircraft systems. This can cause some damage to the cables and equipment within the aircraft electric network. Supply current waveform may be expressed by the Fourier series [11–14]:

$$i_s(t) = I_{DC} + \sum_{n=1}^{\infty} (a_n \cos n\omega t + b_n \sin n\omega t) \tag{6}$$

For three phase 6-pulse diode bridge, the DC output voltage and the RMS input current equal:

$$V_{DC} = \frac{3\sqrt{3}}{\pi} V_m \tag{7}$$

V_m is the maximum phase voltage.

$$I_{RMS} = \frac{\sqrt{6}}{3} I_{DC} \tag{8}$$

Assume losses of the rectifier is zero, therefore the power is

$$P_{out} = P_{in} = V_{DC} I_{DC} = \frac{3\sqrt{3}}{\pi} V_m I_{DC} \tag{9}$$

The input apparent power for the rectifier is:

$$S_{in} = 3 V_{RMS} I_{RMS} = \sqrt{3}\, V_m I_{DC} \tag{10}$$

Therefore:

$$PF = \frac{P_{in}}{S_{in}} = \frac{3}{\pi} = 0.955 \tag{11}$$

Although the power factor is good, the THD value is relatively high and could have a bad effect on the aircraft power systems. The RMS of the input fundamental current for three phase 6-pulse diode rectifier with an inductive load is well known and equals to:

$$I_{1RMS} = \frac{\sqrt{6}}{3} I_{DC} \quad (12)$$

$$THD = \frac{\sqrt{I_{RMS}^2 - I_{1RMS}^2}}{I_{1RMS}} = \frac{\sqrt{\pi^2 - 9}}{3} = 31.08\% \quad (13)$$

The THD could be reduced by using 12-pulse rectifier as shown in **Figure 2**.

12-pulse diode rectifier is fed from a three phase star connected transformer on the primary side, star and delta transformers on the secondary side. Each transformer on the secondary side feeds a three phase 6-pulse rectifier and they add together to form a 12-pulse rectifier, this configuration gives 30° of phase shift which gave harmonics cancellation. The turn ratio of the delta transformer must be multiplied by $\sqrt{3}$ factor in order to get the same voltage level, this illustrated in **Figure 3** [14].

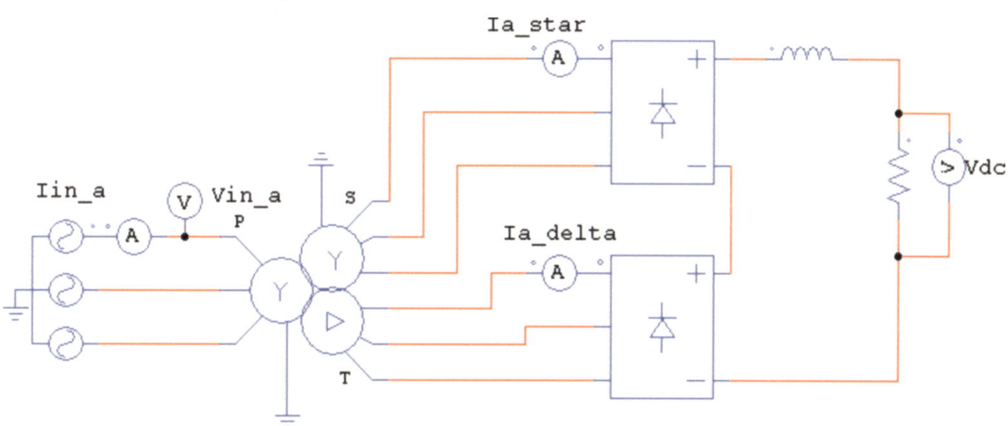

Figure 2. 12-pulse diode rectifier.

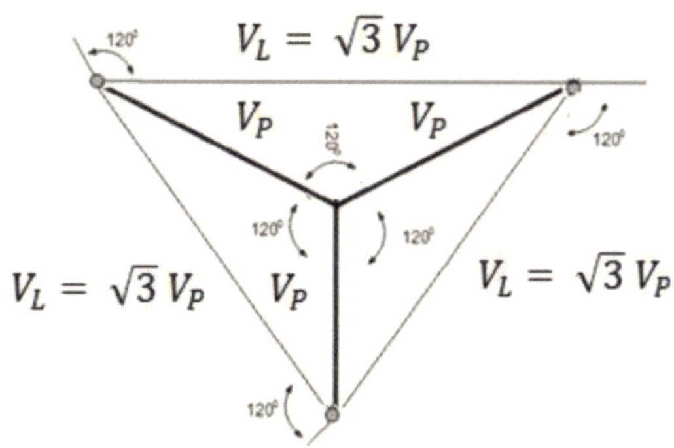

Figure 3. Star-delta configuration.

In a three phase 6-pulse rectifier the dominated harmonics are the 5th and 7th and this why THD is quite high. With a 12 pulse arrangement the 5th and 7th harmonics are canceled as illustrated below [11–14]:

For the star connection, phase (a) current equals:

$$i_{a\_star}(t) = 2\frac{\sqrt{3}}{\pi}I_d\left(\cos\omega t - \frac{1}{5}\cos 5\omega t + \frac{1}{7}\cos 7\omega t - \frac{1}{11}\cos 11\omega t + \dots\right) \tag{14}$$

For the delta connection phase (a) current equals:

$$i_{a\_delta}(t) = 2\frac{\sqrt{3}}{\pi}I_d\left(\cos\omega t + \frac{1}{5}\cos 5\omega t - \frac{1}{7}\cos 7\omega t + \frac{1}{11}\cos 11\omega t + \dots\right) \tag{15}$$

The primary current is equal to the summation of both secondary currents:

$$i_{a\_inp}(t) = 4\frac{\sqrt{3}}{\pi}I_d\left(\cos\omega t - \frac{1}{11}\cos 11\omega t + \frac{1}{13}\cos 13\omega t - \frac{1}{23}\cos 23\omega t + \dots\right) \tag{16}$$

The series has harmonics of an order of $12k \pm 1$ and the harmonics of orders $6k \pm 1$ circulate between the two converter transformers and do not penetrate the aircraft power system network. Since the magnitude of each harmonic is proportional to the reciprocal of the harmonic number, therefore the 12-pulse rectifier has a lower THD equals to:

$$I_{1RMS} = \frac{2\sqrt{6}}{3}I_{DC} \tag{17}$$

$$THD = \frac{\sqrt{I^2_{RMS(12h)} - I^2_{1RMS(12h)}}}{I_{1RMS(12h)}} = 2\,\frac{\sqrt{\pi^2 - 9}}{3} = 15.5\% \tag{18}$$

The THD for the 12-pulse rectifier is reduced by 50% compared with the 6-pulse rectifier.

Figure 4 shows the current waveforms for the supply input current of phase (a) and the currents on the secondary side of each transformer.

Figure 5 shows the harmonics contents for the currents of phase (a) and the currents on the secondary side of each transformer.

For a more THD reduction, 12-pulse active rectifier could be used, this is shown in **Figure 6**. Many advantages are associated with this type of converter:

- The power factor could be controlled by using DQ vectors control.
- THD is very low.
- The rectifier could be operated with a variable input frequency (usually 360–800 Hz) without interrupting its output.
- Bidirectional power flow.

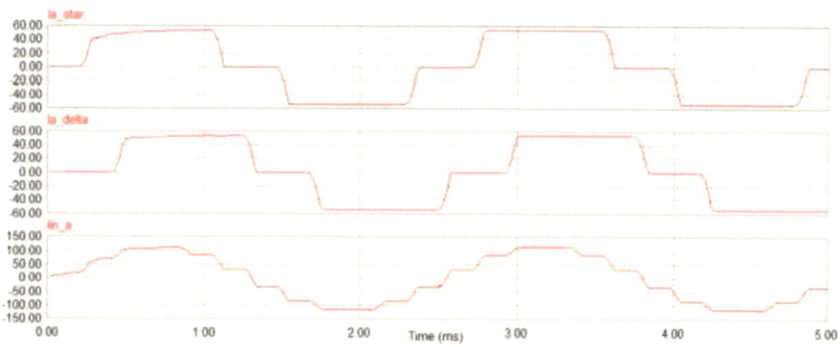

Figure 4. Current waveforms for phase (a) at input frequency 400 Hz.

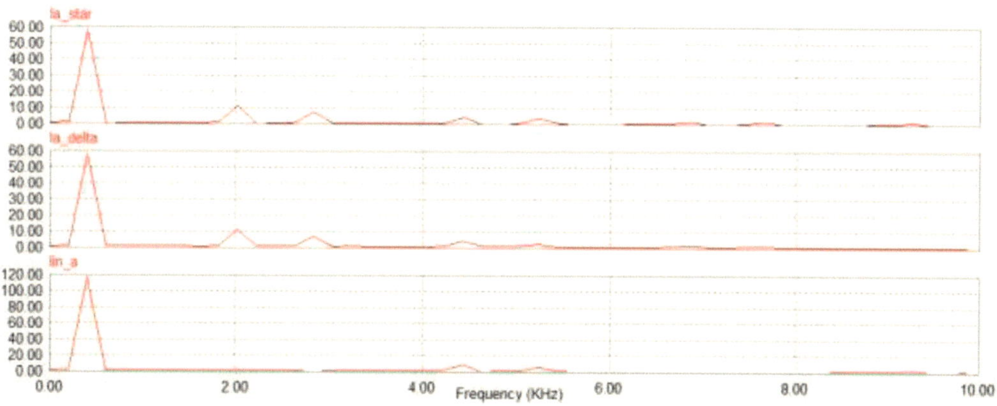

Figure 5. Harmonics contents for the currents of phase (a) at input frequency 400 Hz.

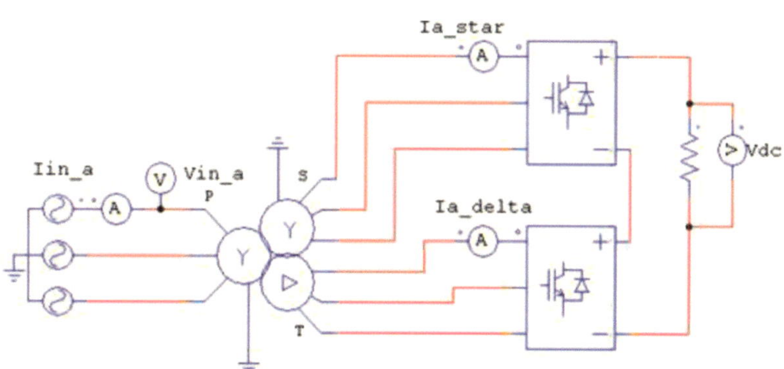

Figure 6. AC/DC 12 pulse boost converter.

3. DQ control circuit

Figure 7 show the configuration of the active rectifier for each secondary side.

The DQ transform is usually called Park transform which is a space vector transformation of the instantaneous 3 phase voltages and currents from a stationary phase coordinate system (ABC) to a rotating coordinate system (DQ) [7, 8].

The general formulas for DQ transformations are given as follows. We assume that the three-phase source voltages v_a, v_b and v_c are balanced and sinusoidal with an angular frequency ω.

The components of the input voltage phasor along the axes of a stationary orthogonal reference frame (α, β) are given by:

$$V_\alpha = \frac{2}{3}V_a - \frac{1}{3}V_b - \frac{1}{3}V_c \tag{19}$$

$$V_\beta = \frac{1}{\sqrt{3}}V_c - \frac{1}{\sqrt{3}}V_b \tag{20}$$

The input voltage can then be transformed to a rotating reference frame DQ chosen with the D axis aligned with the voltage phasor. The voltage components are given by:

$$v_d = V_\alpha \cos \omega t - V_\beta \sin \omega t \tag{21}$$

$$v_q = V_\alpha \sin \omega t + V_\beta \cos \omega t \tag{22}$$

The same transformations are applied to the phase currents:

$$i_d = I_\alpha \cos \omega t - I_\beta \sin \omega t \tag{23}$$

$$i_q = I_\alpha \sin \omega t + I_\beta \cos \omega t \tag{24}$$

Referring to **Figure 7**, let v_{a1}, v_{b1} and v_{c1} be the fundamental voltages per phase at the input of the converter.

$$v_a = Ri_a + L.di_a/dt + v_{a1} \tag{25}$$

$$v_b = Ri_b + L.di_b/dt + v_{b1} \tag{26}$$

$$v_c = Ri_c + L.di_c/dt + v_{c1} \tag{27}$$

where L is the value of input line inductance and R is its resistance of the inductor.

Taking the DQ transformation for the inductor, the input voltage to the converter in the DQ reference frame is given by [15–19]:

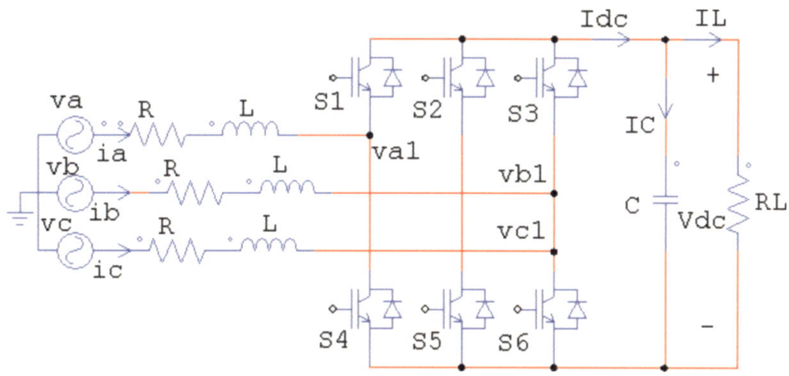

Figure 7. 6-pulse active rectifier configuration.

$$v_d = Ri_d + L.di_d/dt - \omega Li_q + v_{d1} \tag{28}$$

$$v_q = Ri_q + L.di_q/dt + \omega Li_d + v_{q1} \tag{29}$$

Note that v_{d1} and v_{q1} are the DQ components at the converter terminals.

Figure 8 shows the phasor diagrams for DQ coordinates.

The instantaneous active and reactive powers are given by:

$$P_d(t) = 3/2(v_d.i_d + v_q i_q) \tag{30}$$

$$Q_d(t) = 3/2(v_d.i_q + v_q i_d) \tag{31}$$

During the steady state and by assuming the converter losses are negligible, the DC and AC power are equal, therefore:

$$P_d = P_{DC} = V_{DC}.I_{DC} \tag{32}$$

Therefore

$$I_{DC} = \frac{P_d}{V_{DC}} = \frac{3\ (v_d.i_d + v_q i_q)}{2V_{DC}} \tag{33}$$

For a power balance, the delivering power should equal to the absorbing power therefore:

$$P_{AC} + P_{DC} + P_C = 0 \tag{34}$$

Where P_C is the power in the capacitor filter.

If the synchronous frame is aligned to voltage, the quadrature component, v_q = 0. Therefore, the power equations reduce to:

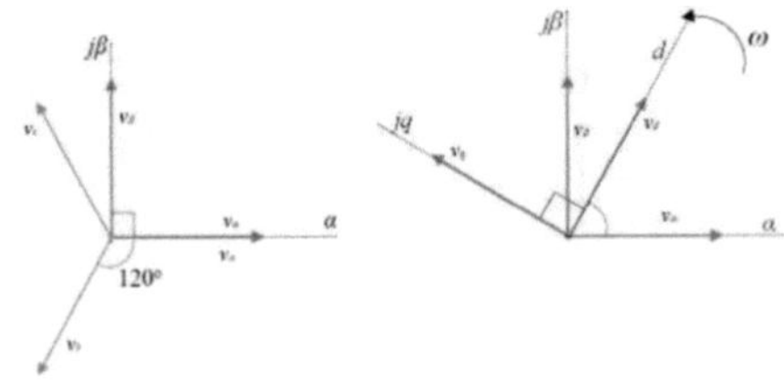

Figure 8. DQ phasor diagrams.

$$P_d = 3/2\ v_d.i_d \tag{35}$$

$$Q_d = 3/2\ v_d.i_q \tag{36}$$

Eq. (32) becomes:

$$I_{DC} = \frac{P_d}{V_{DC}} = \frac{3\ (v_d.i_d)}{2V_{DC}} \tag{37}$$

That gives:

$$P_{AC} + P_{DC} + P_C = 3/2\ v_d.i_d + V_{DC}.I_{DC} + V_{DC}.i_C = 0 \tag{38}$$

Therefore the capacitor current becomes:

$$i_C = -\left(\frac{3\ (v_d.i_d)}{2V_{DC}} + I_{DC}\right) \tag{39}$$

But:

$$i_C = C\frac{dV_{DC}}{dt} \tag{40}$$

From Eqs. (38)–(40):

$$\frac{dV_{DC}}{dt} = \frac{i_C}{C} = \frac{-1}{C} - \left(\frac{3\ (V_d.i_d)}{2V_{DC}} + I_{DC}\right) \tag{41}$$

From Eq. (41) by controlling the active Current i_d the DC output voltage of the rectifier could be controlled.

Inverse DQ transformations then need to be applied to provide the three phase modulating waves (v_{aref}, v_{bref} and v_{cref}) for the PWM generator. DQ vector control has several benefits such as reactive and active power will be easy to control and the dynamic response on the current loop will be very fast.

The PWM generator employs a 20 kHz carrier and is based on a regular asymmetric PWM strategy. The line inductor has a value of 100 μH per phase which limits the Total Harmonic Distortion (THD) to the required value.

http://dx.doi.org/10.5772/intechopen.70882

Figure 9 shows the schematic of the DQ control scheme implemented in the input converter.

The proposed control scheme consists of two parts [13–15]:

1. An outer voltage controller.
2. An inner current controller.

The outer voltage controller regulates the DC link voltage. The error signal is used as an input for the PI voltage controller this provides a reference to the D current of the inner current controller. **Figure 10** shows the DC link model and **Figure 11** shows the close loop control of outer voltage control.

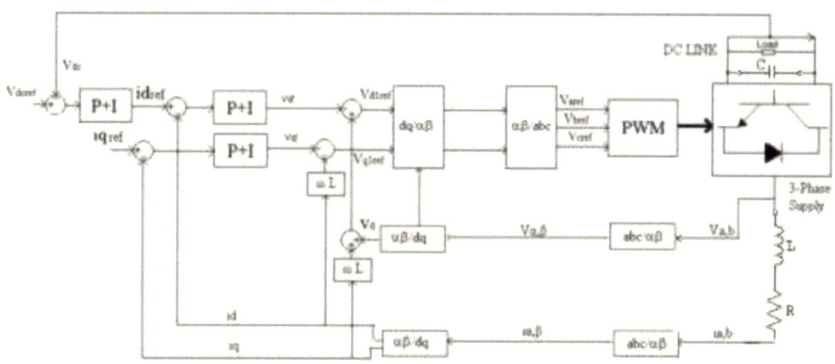

Figure 9. DQ control for the input converter.

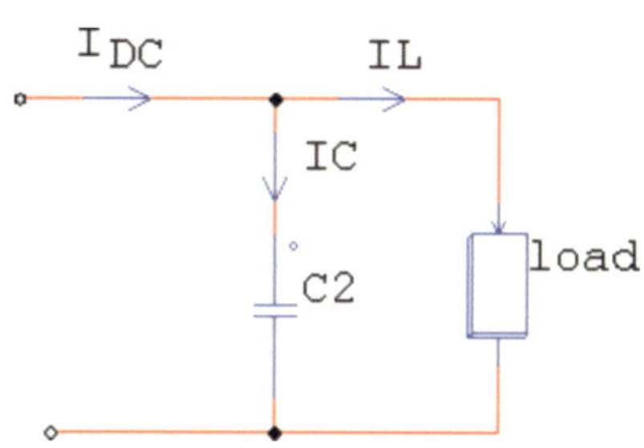

Figure 10. DC link model.

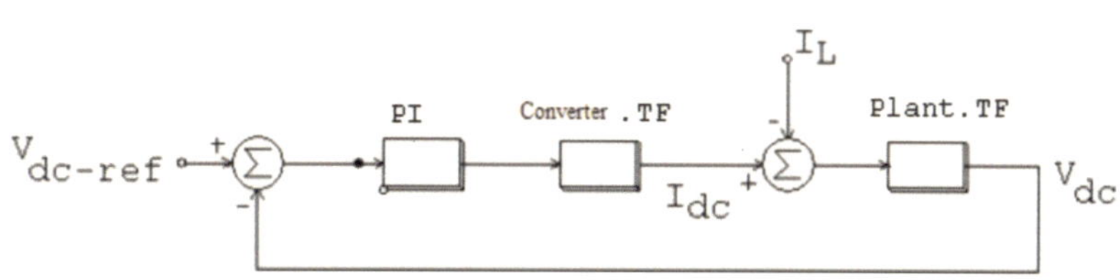

Figure 11. Close loop control of outer DC voltage control. PI TF: $K_p + K_I\frac{1}{S} = K_P\left(\frac{S+a_i}{S}\right)$; converter TF: $\frac{3}{2\sqrt{2}}M$; plant TF: $\frac{1}{CS}$.

The relationship between the DC voltage and the D axis input voltage is given by:

$$V_{DC} = \frac{2\sqrt{2}V_d}{M} \text{ or } M = \frac{2\sqrt{2}V_d}{V_{DC}} \tag{42}$$

$$I_{DC} = \frac{3}{2\sqrt{2}} M I_d \tag{43}$$

Where: M is the modulation index.

For PI controller:

$$\text{PI TF} = K_p + K_i \frac{1}{S} = K_P \left(\frac{S + a_i}{S} \right) \tag{44}$$

Where $a_i = \frac{K_i}{K_p}$

For the converter the system has the following TF:

$$\text{Converter TF} = \frac{I_{DC}}{I_d} = \frac{3}{2\sqrt{2}} M \tag{45}$$

For the plant TF, The dc link may be modeled by a capacitor:

$$\text{Plant TF} = \frac{1}{CS} \tag{46}$$

Therefore the characteristic equation for the DC link voltage control is given by:

$$S^2 + \frac{3MK_p}{2\sqrt{2}C} S + \frac{3MK_p a_i}{2\sqrt{2}C} = 0 \tag{47}$$

General equation for second order characteristic equation is given by:

$$S^2 + 2\xi\omega_n \, S + \omega_n^2 = 0 \tag{48}$$

Therefore the controller parameters are given by:

$$K_p = \frac{4\sqrt{2}\, C\xi\omega_n}{3M} \tag{49}$$

$$a_i = \frac{2\sqrt{2}C\omega_n^2}{3MK_p} \tag{50}$$

where ω_n and ξ are the closed loop natural frequency and damping ratio, therefore the controller parameters can be easy calculated by choosing the value of the modulation index, ω_n and ξ.

http://dx.doi.org/10.5772/intechopen.70882

A PI inner DQ current control refers the phase current measurements to a rotating co-ordinate frame DQ fixed to the supply voltage. **Figure 12** shows the DQ model of input stage and **Figure 13** shows the close loop control of the inner current control. If the phase currents are in phase with the supply voltages, the current referred to the direct D axis becomes the DC link current and the current referred to the quadrature Q axis is equal to zero. The co-ordinate transformation is done using phase angle information derived from the measurement of the supply voltages. However, if the system is needed to operate with a leading or a lagging power factor, the Q axis reference value could be changed to define the displacement angle of the rectifier. The D axis and Q axis currents are compared to their respective demands values and the error is applied to individual PI controllers give voltage demands referred to D axis and Q axis. In the rotating co-ordinate frame the D axis and Q axis currents are inter-related due to their rotation. The rotation introduces an orthogonal component into time derivative of each current which, when applied to an inductive load, gives a voltage components along the axis orthogonal to that of the current. The DQ scheme studied uses two feed forward terms:

- $\omega L i_q$ fed into the V_d demand.

- $\omega L i_d$ fed into the V_q demand.

These feed forward terms de-couple the two currents. In addition the supply voltage is referred to D axis and this added to the V_d demand to avoid the integrator having to compensate for it. The final demand voltages are transferred back into stationary co-ordinates and the resulting sinusoids are used to generate the PWM.

Figure 13 shows the close loop control of inner current control. The following transfer functions are applied to the control block.

The plant is represent the line from the generator to the input of the converter which has RL network with following TF:

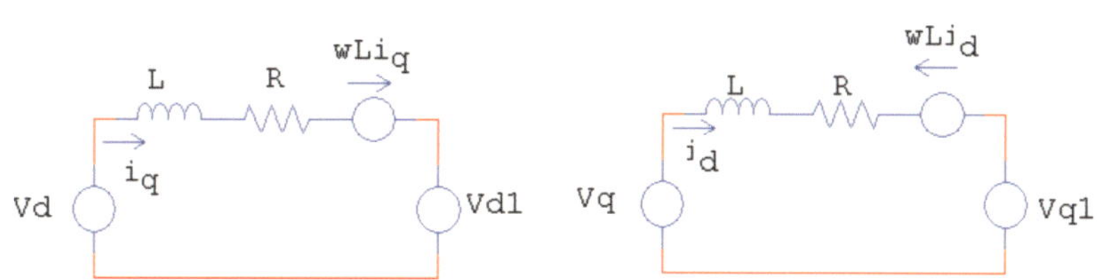

Figure 12. Equivalent circuit for DQ input supply.

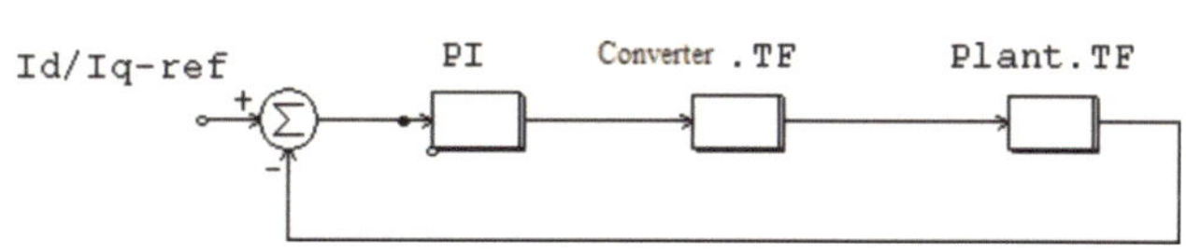

Figure 13. Close loop control of inner current control. PI TF: $K_p + K_i\frac{1}{S} = K_P\left(\frac{S+a_i}{S}\right)$; converter TF: $\frac{1}{1+TS}$; plant TF: $\frac{1}{R+LS}$.

$$\text{Plant TF} = \frac{1}{R + LS} \tag{51}$$

The converter may be modeled as a first order lag. $= \frac{1}{1+TS}$ where $T = \frac{1}{2F_s}$

F_s is the switching frequency. The same procedure can be used to calculate the parameters of the controller. Simulink or other tools can be easily used to tune the PI controller by using Zeigler-Nichol's method.

4. Simulation results for the 12-pulse active rectifier

In order to optimize the power quality and transient behavior of the power distribution system, a well-designed simulation model of the 12-pulse active rectifier based on detailed component models will be necessary.

For high voltage demand, the two rectifiers are connected in series, and for high current demand, the rectifiers may connect in parallel. The converter has been simulated for various operating conditions with the following parameters.

- Input inductance for $L = 100\ \mu H$, input resistance 0.2 Ω. For each converter
- DC filter $C = 200\ \mu F$. For each converter
- Switching frequency 20 kHz. For each converter
- Input frequency 360–800 Hz.
- AC input voltage = 115 V RMS.
- Resistive load = 15 Ω.
- The DC voltage reference for each converter is set to 320 V.

Simulation results show that comparing with a conventional 12-pulse diode rectifier, the low order harmonics are totally eliminated and only very low harmonics around the switching frequency at frequencies $f = mf_s$ where $m = 1, 2, \ldots, \infty$

Figures 14–19 show different simulation results.

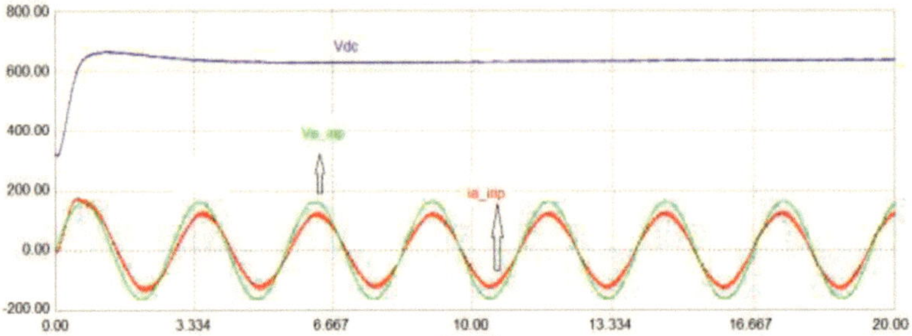

Figure 14. Waveforms results for 360 Hz input frequency.

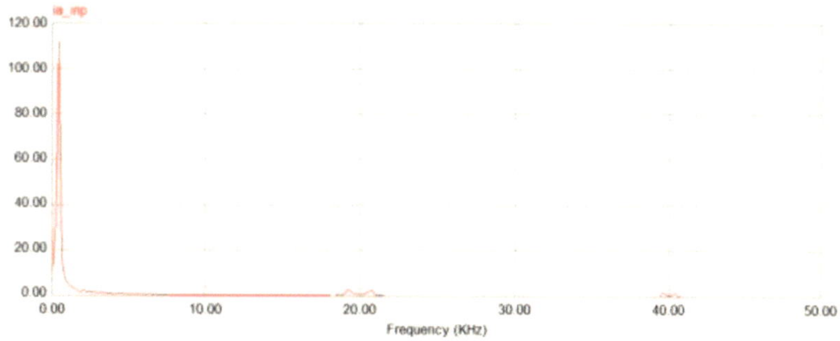

Figure 15. THD for phase a current—input frequency 360 Hz.

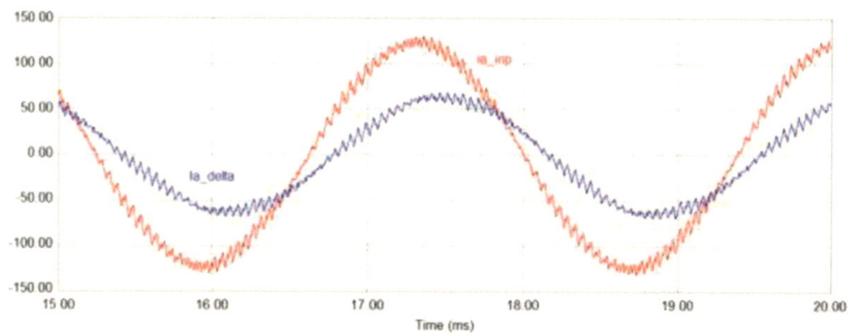

Figure 16. The 30° shift for the delta current and the input current for phase a—input frequency 360 Hz.

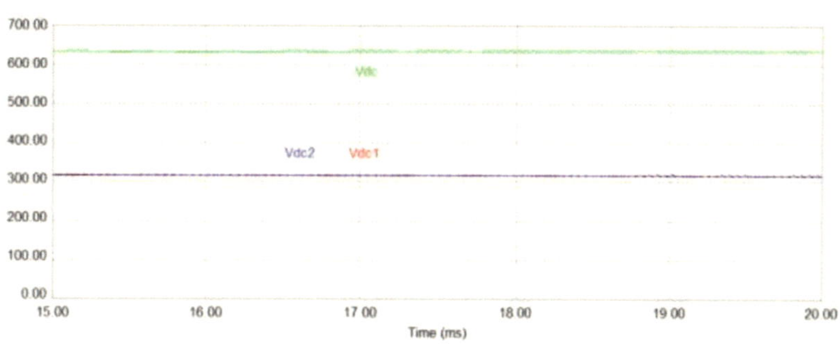

Figure 17. DC voltage level for each converter and the overall DC voltage.

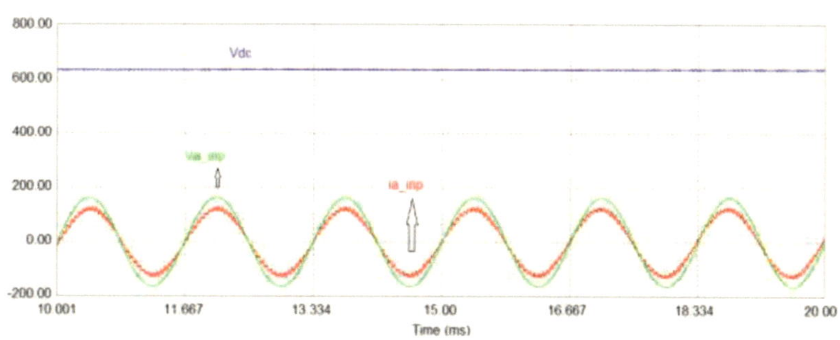

Figure 18. Waveforms results for 600 Hz input frequency.

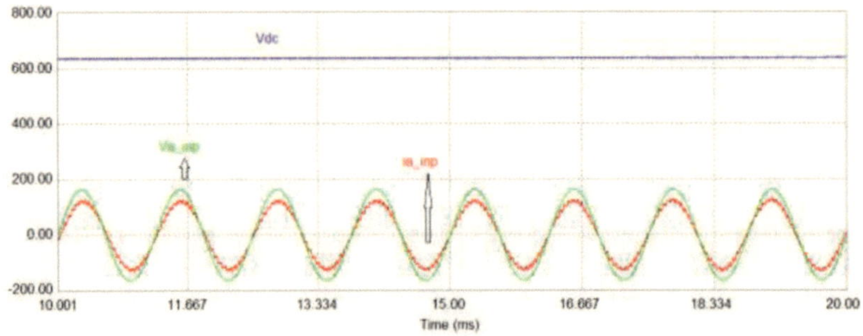

Figure 19. Waveforms results for 800 Hz input frequency.

5. Conclusions

With the future use of advanced power electronics the 12-pulse active rectifier gives a beneficial approach within the aircraft power distribution system. Any Low frequency current harmonics could be eliminated and there is the possibility to operate the rectifier at a variable power factor in order to provide system level benefits. Furthermore it has been shown that; there is the possibility to operate the system with variable input frequencies while keeping the input current harmonics low.

Acknowledgements

I would like to express my sincere appreciation and respect to the late Prime Minister Rafik Hariri who supported my studies in England.

Author details

Mohamad Taha

Address all correspondence to: tahamh@rhu.edu.lb

Rafik Hariri University, Lebanon

References

[1] Faleiro L. Beyond the more electric aircraft. Aerospace America by the American Institute of Aeronautics and Astronautics. September 2005:35-40

[2] Rosero JA, Orteg JA, Aldaba E, Romera L. Moving towards a more electric aircraft. IEEE A&E Systems Magazine. March 2007:3-9

[3] Ahistrom K, Torin J. Future architecture of flight control systems. IEEE A&E Systems Magazine. Dec 2002;**17**(12):21-27

[4] Faleiro L. Initial research towards a more electrical aircraft. In: More Electrical Aircraft Conference Royal Aeronautics Society; 2004. pp. 245-257

[5] Wheeler P, Bozhko S. The More Electric Aircraft: Technology and challenges. IEEE Electrification Magazine. December. 2014;**2**(4):6-12

[6] Sarlioglu B, Morris CT. More electric aircraft: Review, challenges, and opportunities for commercial transport aircraft. IEEE Transactions on Transportation Electrification. June 2015;**1**(1):54-64

[7] Taha MH. Active rectifier using DQ vector control for aircraft power system. In: IEMDC; 2007. pp. 1306-1310

[8] Taha MH, Skinne D, Gami S, Holme M, Raimondi G. Variable frequency to constant frequency converter (VFCF) for aircraft applications. In: PEMD; 2002. pp. 235-240

[9] Guerreiro J, Pomilio A, Curi Busarello TD. Design and implementation of a multilevel active power filter for more electric aircraft variable frequency systems. In: Power Electronics Conference, Brazilian; 2013. pp. 1001-1007

[10] Corcau JI, Dinca L. On using PEMFC for electrical power generation on more electric aircraft. International Journal of Electriconics and Electrical Engineering. 2012;**6**(2):186-189

[11] Prachi G. Effect of harmonics on active power flow and apparent power in the power system. IOSR Journal of Electronics and Communication Engineering (IOSR-JECE):39-43. ISSN: 2278-2834, ISBN: 2278-8735

[12] Almoataz YA. Sources and mitigation in harmonics in industrial electrical power systems: State of the art. The Online Journal on Power and Energy Engineering (OJPEE). 2012;**3**(4):320-333

[13] Sangeeta Sarali D, Shailaja P. Mitigation of harmonics using thyristor based 12 pulse voltage source PWM rectifier. IJRET. Nov 2012;**1**(3):267-271

[14] Gong G, Drofenik U, Kolar JW. 12-pulse rectifier for more electric aircraft applications. In: IEEE International Conference on Conference: Industrial Technology; Dec 2003;**2**: 10-12

[15] Faheme K, Chariac D, Sbita L. Control of three phase voltage source PWM rectifier. In: 3rd International Conference on Automation, Control, Engineering and Computer Science ACECS 16, Proceeding of Engineering and Technology (PET); 2016. pp. 649-654

[16] Habetler TG. A space vector-based rectifier regulator for AC/DC/AC converters. IEEE Transactions on Power Electronics. 1993;**8**:30-36

[17] Eid A, Abdel-Salam M, El-Kishky H, El-Mohandes T. Simulation and transient analysis of conventional and advanced aircraft electric power systems with harmonic mitigation. Electric Power Systems Research. Apr. 2009;**79**(4):660-668

[18] Biagini V, Odavic M, Zanchetta P, Degano M, Bolognesi P. Improved dead of a shunt active filter for aircraft power systems. In: Proc. IEEE ISIE; Jul. 4-7, 2010. pp. 2702-2707

[19] Kazmierkowski MP, Dzeiniakowski MA, Sulkowski W. Novel space vector based current control for PWM inverters. IEEE Transactions on Power Electronics. 1991;**6**:158-166